I0481023

HOUSE ELECTRICAL DIY THE BOOK OF NEWBIE'S

Fully Updated House Electrical Circuits and Light Designs/ Estimations, Backup Power and Also Requirements for Beginners.

Michael H. Maurice
Michael H. Maurice Associates Inc

Copyright Notice:

All designs and recordings are intellectual property of **Michael H. Maurice Associates Inc.**, as well as any type of unauthorized usage or recreation without consent is restricted.

This presentation is secured by UNITED STATE and International copyright regulations. Reproduction and distribution of the discussion without created approval of the en-roller is prohibited.

Author Details:

Michael H. Maurice identified the consumer demands, developing a leading and also trusted on source for the best details, inspiration, as well as guideline related to the house and also home. He is the leading publisher of books on all elements of embellishing and style; house repair and enhancement; residence plans; gardening and also landscape design; and grilling.

Michael H. Maurice's publications is unique from various other publications with their complete and easy to follow directions, current information, and also extensive use of shade photography. Amongst its very successful title is *HOUSE ELECTRICAL DIY THE BOOK OF NEWBIE'S.*

Michael H. Maurice's also have emphasis of style and also development on the Electronic devices, Physics as well as Mathematics. He was an Engineering Graduated that has actually headed his own firm, Michael H. Maurice's Associates Inc., since 2015. He worked in the IT Company which publish the electronic books publication for the Engineering Pupils. But he wished to start it by individually for publishing the book for education function. "Before starting his independent consultancy, he held a several placements at the US consisting of purchases editor, supervisor of electronic publishing. He holds engineering levels from MIT. In addition to his modified reference works, he is the author of four books.

Legal Disclaimer

CAUTION: Specific house renovation jobs are inherently dangerous, as well as also the most benign tool can create significant injury or fatality if not used appropriately. ALWAYS READ AND FOLLOW USER'S MANUAL As Well As SAFETY AND SECURITY WARNINGS. You need to be particularly mindful when dealing with electricity always use common sense.

Any recommendations, assistance or other info provided on this book or within any one of our publications cannot totally expect your situation. If you go to all unclear regarding finishing any kind of facet of this or various other house wiring jobs, consult a competent electric service provider to perform the service(s) for you.

ALWAYS adhere to electric code needs particular to your area, as well as before embarking on any type of home electric task, contact your local electric authority and your insurance provider to guarantee that you abide by all plans, service warranties, policies and also authorities concerning this job.

No recommendations or info, whether oral or written, acquired by you from us or with the solution, it's staff members, specialists and/or professionals shall produce any kind of guarantee not expressly made here.

By utilizing this book, consisting of any kind of applets, software application and content consisted of therein, the reader concurs that using this book and also its details item is completely at his/her very own risk.

Index

Introduction

Know about your house electrical is the key subject in the house. As we know, every time we are hiring an experts for these jobs. This book will help you in all aspects like planning, developing and safe execution, even if you hire an expert for your every work at your home or garden area, guest house, etc. As we conveyed in this book, may be the value in completing your house projects estimation and cost-effective.

Electrical basics, selection of materials, and execution are major chapters in this book. We hope, the enough strong base available in these book to execute jobs in safe.

However, we are recommending the executer must be minimum knowledge of electrical with few years' experience in the field of electrical.

Owner's responsibility increasing at new projects

Initially, when you think about the house modification or expanding your house electrical facilities then hiring expert and make a best estimation and plans. All basic plans are made by your house requirements and budgets. Knowing the professional's quality is one of the important things and their crew members. The owner's still monitoring the plans and move up to forward progress on the job.

In the modern life, Electrical utility designs are much advanced and reduce the people time. All the new assignment plans are must to be get approval from the local government and it need to meet out the Electrical Standards like IEEE, IEC, NEC and etc. This is too depends where you living do and circumstance.

Why need to be involved in the Electrical Work?

1. To maintain the system for a long time
2. Save money
3. Increase the work efficiency
4. Follow the Electrical standards
5. Future developments

The book mentioned that all standards which is mandatory to the House electrical system.

SAFETY
Hazards and Control measurements into your work

Electrical is one of the strongest type hazards and the injuries can't be tolerate. There are many chances to get electrical leaks in the house circuits like in the garden area, main board area, playing area and stair case area and etc. The small current is enough to lead a heavy shock and injuries because the voltage pressure is always will be high in the system.

The book is giving enough knowledge about the electrical hazards and fire safety. The concern is familiarizing to safe execution who's going to work in the electrical.

On upcoming sections will explain about several kinds of Hazards and preventions.

1. Arc Faults

1.1. Know about the Arc Faults and Flash

The reason for the more fatalities is Electrical fire than Electrical shock, cause inhalation of smoke. Electrical fires begin with series and parallel arc faults and less quality or less capacity of the conductors in the conduits pipe.
These kinds of installation mistake and improper design are major role to damage or worse injury or some worst cause lead to death.

Arc faults are usually initiates, when increasing the heavy loads (except the starting loads) like motor running, heating devices, Air conditioners, geezers, etc. with inefficient size of the conductors or less quality wires.

The other interesting reasons, insufficient rated fuses, low quality protection materials such as MCB, Links, Overload relay to trip out. So these faults could be cleared when the fault can be isolated from the circuit (like switching off the main source).

1.2. How to prevent form the Arc faults

1. Select the sufficient size conductors and quality.
2. Tightness of the Cable terminations
3. Conduit pipe and wall pipes proper fix
4. Select the suitable protection device such as fuse, relays, MCB.
5. Proper Operations and maintenance.
6. Prefer the high quality Switches and Junctions box(Preference is metal box so that avoid building fire) refer image-1
7. Fixing fire detector and fire alarm switches

Power cable termination at Main distribution Board

2. Electrical Shock

What is an electric shock?

The other strongest Hazards is electrical shock and leaks, when human touch on the live electrical terminals such as plug socket, open wires or some defective material which have al leak in the body. The current it's passes through the human body to earth or the other co-phases.

This is shock is enough to make internal and external damages like tissue and organs.

What are the sources for electrical shock?

- Power Circuits
- Home appliance
- Pump motors or Blenders
- electric Tools like grinding, Drilling machines
- electrical outlets like sockets or temporary extensions

Due to these shock are bit less severe but it can become fast transfer to serious when the person holding a longtime, also if kids catching or put their mouth from the outlet cables.

2.1. How to Avoid Electrical shocks?

Whenever, start the electrical work we should clearly know about the job procedures. The job procedure and checklist will help to execute the clear job. So when the above cleared then shock risk will be minimal.

A frequent chance to shock is using wrong equipments in the power circuits. Standards are mentioning that insulted tools must be handled while online or offline work in the electrical.

Proper communications between the people when switching ON/OFF happening and miscommunication will lead to severe penalties in the human health.

2.2. What are Symptoms of Shock?

Potentials and current is playing major role that how person getting severe injure. The below are few symptoms of the shock.

- Skin burn
- spasms of muscle
- Conscious lose
- Hard Breathing
- headache
- Vision and hearing issue
- Heart beat irregular or stop

2.3. What should person do when some got electric shock?

To avoid the big impact of the human health, some immediate response necessary to taken.

If the person got severely shocked then what can be done as follows,

- Cut off the electrical source from him to avoid the flow of the current.
- When the person get a serious attack, if lost the heart beats, isolate the person from the electrical and give a proper Cardiopulmonary resuscitation (CPR). Note: The CPR shall be avoided when person breathing is even bit OK.
- Call the Emergency number 911 or local emergency number to call the medical facilities at immediately.
- Don't move the person unless the power is OFF, else source will catch the 2nd person also.

Minor shocks

- Give the medical facilities and mental counseling as soon you can, it's better to take one complete medical checks even the person is seems normal.
- If slightly gets skin 1st or 2nd level burning then give a proper first aid and put on lotion for heal.

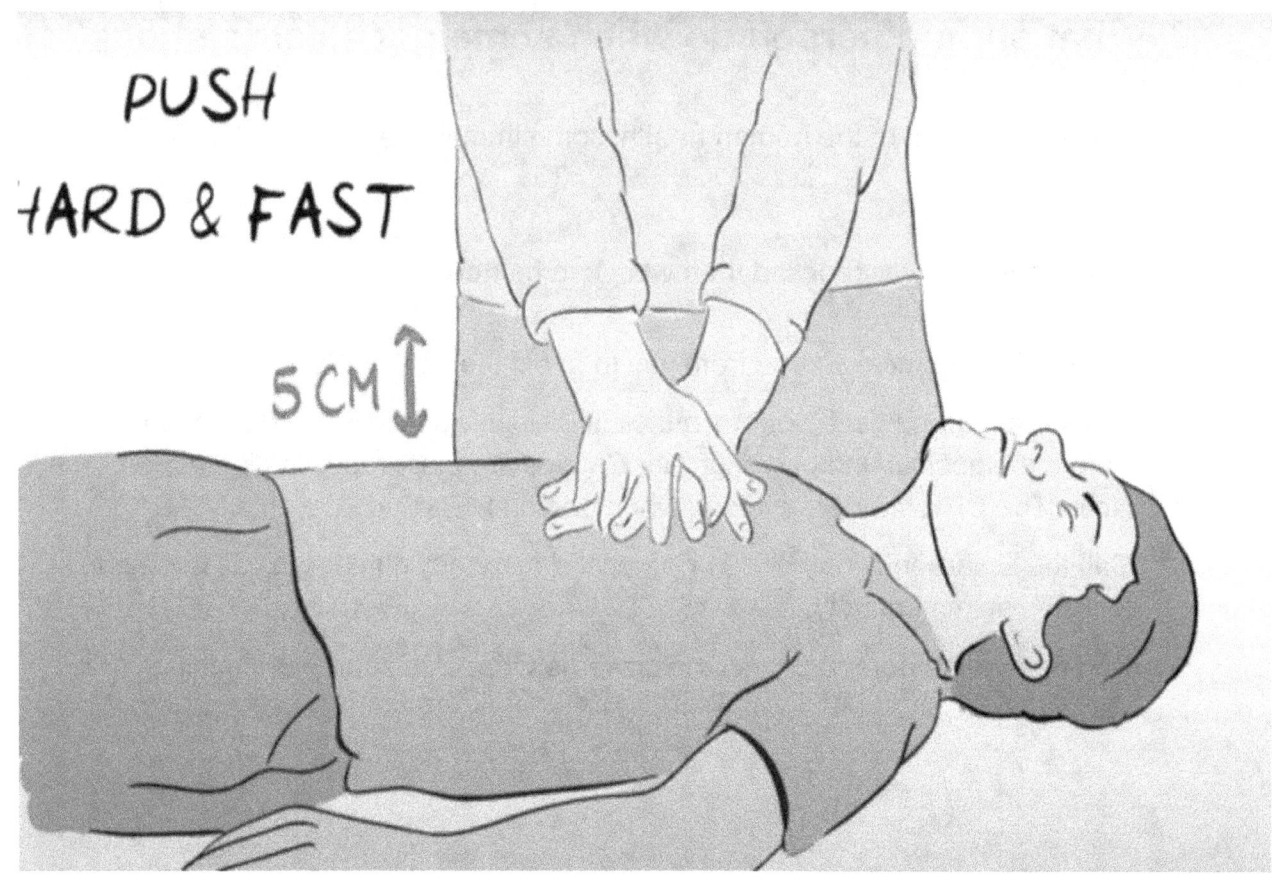

2.4. Know about long term effect?

There is chance to lasting memory power and permanent skin scare marks, get mental effect when the person get the shock his head and brain area.

Some shocks will be pain, tingling and muscle weakness because of inside attacks.

3. Reference-Multinational Electrical Safety

Here, Multinational Electrical and Fire Safety Standards (source reference from Wikipidea)

- United States (USA) NFPA, IEEE STD 80, IEEE STD 80 - United States NFPA 496,NFPA 70
- India (IND) India Standardization - India - IS-5216,IS-5571,IS-6665
- China (CHN) - China GB4943, GB17625, GB9254
- Brazil (BRA) Brazilian National Regulation - NR10 Brazil
- Poland (POL) Polska Norma - Poland - PN-EN 61010-2-201:2013-12E
- Bulgaria (BUL) Български-Държавен-Стандарт- БДС 12.2.096:1986
- Australia (AUS) Australian Standards - Australia - AS/NZS 3000:2007,AS/NZS 3012:2010,AS/NZS 3017:2007,AS/NZS 3760:2010,AS/NZS 4836:2011
- Germany (GER) IEEE/TÜV - Germany NSR 2014/35/EU
- France (FRA) La norme français C 15-100 - Aspects de la norme d'installation électrique France
- Great Britain (GBR) British standard - United Kingdom[16] BS EN 61439,BS 5266,BS 5839,BS 6423,BS 6626,BS EN 62305,BS EN 60529
- Russia (RUS) - ГОСТ Р 52726-2007, ГОСТ 12.2.007.0-75, ГОСТ 12.2.007.0-75, ГОСТ Р МЭК 61140-2000.

4. Lightning Protection Standards

- China (CHN) GB/T 36490-2018
- Russia (RUS) СТО 083-004-2010,ГОСТ Р МЭК 62561.2-2014
- Bulgaria (BUL) БДС EN 62305-1:2011
- Germany (GER) DIN EN 62305-1
- United States (USA) NFPA 780, IEC 62305
- France (FRA) Norme NF C 15-100
- Poland (POL) PN-EN 62305
- India (IND) IS 2309
- Great Britain (GBR) BS-EN 62305
- Spain (ESP) UNE 21186. Protection contra el rayo

Basic Electrical Concepts

Electricity, it's a kind of natural energy and actually invented by nature, many misconceptions abound to who discovered it. In this modern world, all lives depend on the electrical, without electrical nothing will be effectively saved even our culture and survival. Electrical is playing a big part in the medical field, "Could you image, how many lives are getting saved with help of the Electricity in the medical field" in a single day. Electricity is a stressbuster and helping to maintain a stable mind for a human by a lot of entertainment, social media and especially in hashtags and TV shows.

An Ancient Greeks were discovered an amber rubbing fur caused on with attract of two that we can call as static electricity. Even 1930's research discovered a pot inside the copper sheets which might be used as an Electrical battery at roman sites.

But on the 17th century, some research said electrical related things were discovered such as electrostatic generator and classification of the positive and negative currents, conductors and insulators.

Many credits to Mr. Benjamin Franklin who had done the more research about the Electricity, his tests and experiments were helped to invent the lightning.

On 1752, Franklin done a many research about the electrical lightening, Also he did an experiment on the Kite, key and storms.

Mr.Benjamin Franklin Image source: Wikipedia

1. What is Electricity?

In generally movement of the discharge where it's a positive to negative charges, when carpet and door knob are a bad combination get an electric charge by friction and no matter how it's getting charged, When convention to be positive to negative, batteries are chemically and frictions are physically(Door and Carpet).

2. What is Current?

The rate of flow of electrical charge which can move through the metallic conductor such as copper or aluminum. There are two types of the current,

1. AC (Alternative Current)
2. DC (Direct Current)

AC current is changing the direction with the time cycle which known as frequency along with sign of phase and neutral voltage.

DC current flow in the same direction certain constant voltage with sign of positive and negative. This DC circuits are not much suitable for the long line transmission.

Noticeable history had written by Thomas Edison and Allessandro Volta about DC circuit in Electrical. Nikola Tesla found an alternate current for a long line transmission later.

AC Current is possible to produce in multilevel for easy transmission where DC is very limit. But DC current widely used in the portable device such as trimmer, dryer, portable lights and etc.

Generally, House Electrical side we are using the AC and DC circuits. AC power receiving from the electricity board and most of the appliances are run on AC circuits. Since, DC circuits are small voltages which can be received from the battery source and AC-DC chargers.

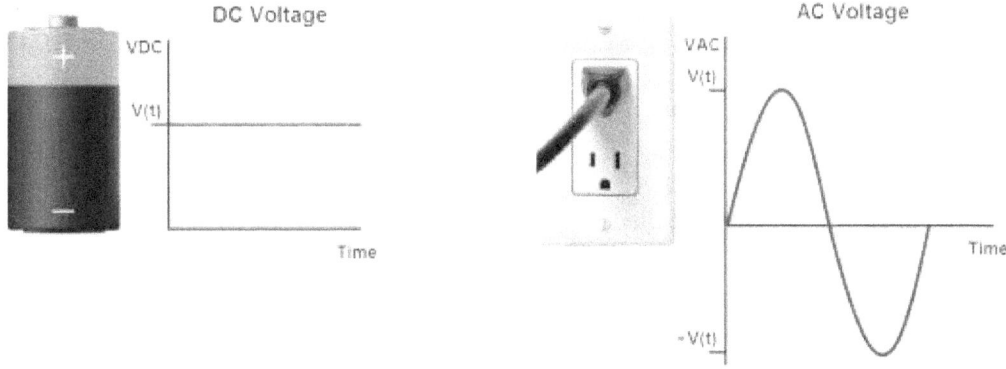

3. What is Voltage?

Voltage is the pressure to push the electrical current to enabling the close circuits through the load. Voltage, provided by energy such as battery or ac circuits which can power up the light or any load.

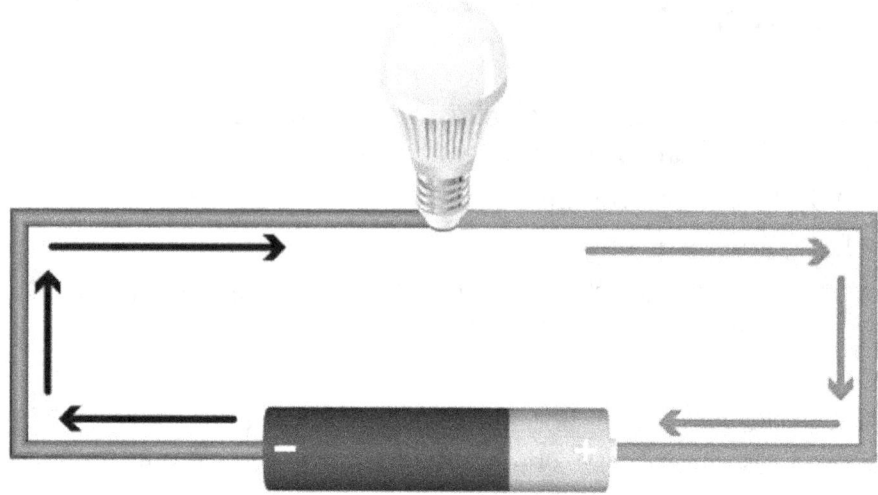

Voltage can be either Direct Voltage or Alternate voltage. Direct voltage is same polarity all the time, where Alternate is having a reverse and forward polarities direction with periodic time cycle

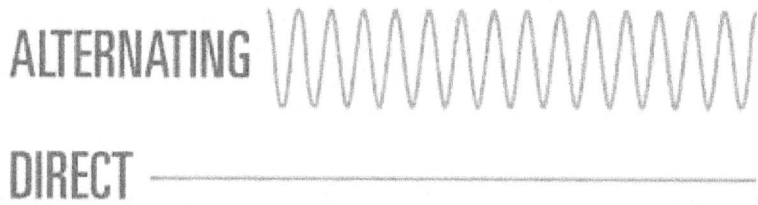

The number of cycles is known as frequency, which known as Hz in unit. Voltage produces an electrostatic field, even a no current moving in the circuit. Usually, No voltage will increase or decrease with two specific point with the distance. The density of diminishes in the zone between the two points which known as Phase and Neutral.

On AC circuits, Voltage and current are total three balanced angle with positive and negative directions. As figure shown here, This phase angle will be applicable for the both current and voltage.

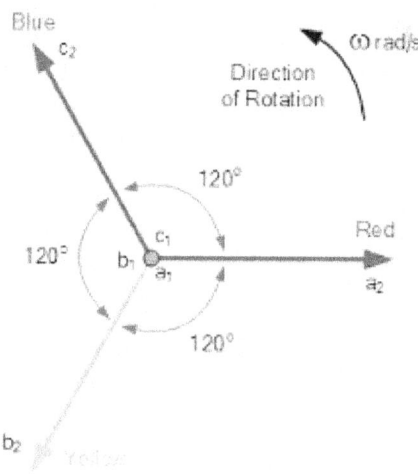

This phase split will 120 degree at theoretically but the due to the characteristic of the load it can be slightly increase or decrease with in the phase.

In domestic loads, single and three phase circuits utilizations are available.

3.1. Available Voltage Levels in Domestic and Industries

On this topic, we have multi range of levels available in the entire world. We are giving some general information about the voltage level and their frequency range. There are nearly forty countries are operating in sixty hertz frequency and rest of the countries are operating in fifty hertz.

Single phase supply is the primary supply for the domestic utilities such as house, shop and hotel. Where, 3 phases are operating in some loads in house or hotel such as pump motors or special utilities, Most of the industries are operating by 3 phase sources because of the heavy loads such as Motors and machines.

Here, we have mentioned the voltage range based on the above requirements along with frequency. This below is just for the information purpose, if the home owner is going to work with only single phase then consider the single phase details are enough.

Phase Reference	Voltage	Frequency	No. of wires
Phase - Neutral	110-120V	60Hz	2 wires
Phase - Neutral	230V	60Hz	2 wires
3 Phase (R-Y-B)	220V	60Hz	3wires/ 3wires and Neutral
3 Phase (R-Y-B)	380-400V	60Hz	3wires/ 3wires and Neutral
3 Phase (R-Y-B)	415-440V	60Hz	3wires/ 3wires and Neutral
Phase - Neutral	110-120V	50Hz	2 wires
Phase - Neutral	230V	50Hz	2 wires
3 Phase (R-Y-B)	220V	50Hz	3wires/ 3wires and Neutral
3 Phase (R-Y-B)	380-400V	50Hz	3wires/ 3wires and Neutral
3 Phase (R-Y-B)	415-440V	50Hz	3wires/ 3wires and Neutral

The number of wires and sockets & plug used different types by region. The each countries are developing their electrical grids in multi range of rate due to their countries ambient temperature and the soil nature.

4. What is the Electric Power?

Electric power, Energy is the one form to another form conversion such as mechanical energy to electrical energy and again electrical energy to convert to mechanical energy conversion, chemical, computer, lighting, entertainment and many other industries use.

In term technical of Electrical power SI unit is Watt power one joule per second.

E=V x I

Where
E= Energy (Power)
V= Voltage
I= Current (Ampere)

Energy is consuming, when current flows through the load. The load can be resistor, lights, fans or any other electrical loads.

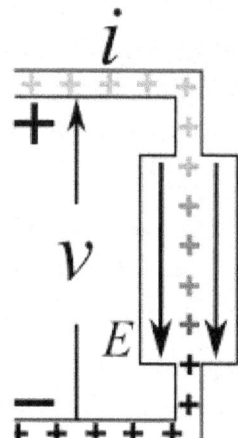

The concept is used for the Heaters, running equipment like motors, fan, heating stove, microwave ovens, TV, fridge and all other loads. However, the current flow creates the heat and it's dissipated the electrical power.

This power is calculating as Watts, Kilowatts and Megawatts depends on the various loads. Based on the electrical power is converted as unit consumptions.

We will see the small example of Power Calculation for the room light.

Let's take a 100 Watts light

Rated Voltage for the light=230V

Current will be 0.434 Amps

E = V x I

100 = 230 x 0.434.

Know about the Earthing System and Uses

Over all, in every electrical installation made, earthing is most important factor. The earthing system for this reason manages the conductors with respect to the ground's conducting surface.

What's the preference of an Earthing?

Earthing is a mandatory element of electric systems because of the sticking to factors:

1. It stops damages to electric devices and likewise devices by preventing heavy faulty current from going through the circuit
2. It stops the threat of fire that can or else be caused by leakage current
3. It keeps individuals risk-free by stopping electric shocks

1. Benefits of Earthing

On the technological perspective, earthing has some superior benefits, resulting it in becoming a mainstream technique in the electric sector such as protection and fault finding medium. One more advantage is that metal can be utilized in electric arrangements without needing to fret concerning conductivity. Metal is a wonderful conductor of electricity, right earthing guarantees that steel elements not indicated to be utilized for Current-Transformer can be contained in the system.

This is done by providing a separate line for this defective existing, enabling its timely discovery as well as also trip and isolate the faults.

The electric system is linked to the general earth mass as well as additionally avoid to reach a different voltage. The Voltage of the earthing is absolutely no volts as well as additionally is referred to as the neutral of the electric energy supply. This aids in keeping the equilibrium to the circuits. In cases of rises in the voltage, high voltages can experience the electrical power circuit. These kinds of overload can bring about damaging of devices as well as threat to human life. The existing is sent using a various program as well as also does not influence the electrical system when earthing is set up with the electric configurations.

2. Types of Earthing system.

2.1. Earthing used by pipes

A galvanized iron pipeline is selected ought to be such that it has openings punctured at regular gap and the pipe shall be slim at the bottom end.

A clamp is connected to the G.I pipe to which an earth wire is connected. This pipeline diverts the electric conductance inside the earth.

The pipeline is put in the earth pit at deepness not less than 3 meters. The room inside the G.I pipeline is filled by the alternate layers of salt and charcoal as much as the clamp level.

A galvanized steel as well as a pipe that has openings at regular periods are maintained inside the planet. Keeping in view its affordable Pipe earthing is typically used for all domestic purposes.

Water is poured into the G.I pipe to preserve earthing resistance within the defined limits. This can be done by open the chamber brick maintenance cover and approach the pipe mouth to fill the water.

2.2 Earthing used by plate

The plate product is of either copper or galvanized iron is used in the plate earthing system. This plate chosen ought to of certain defined measurements, which is placed inside the earth at a depth less than 3 meters from all-time low. This plate is linked to the electrical conductors to draw away the electrical cost inside the earth. The layout of Plate Earthing is offered below.

2.3. Earthing used by Rods

Rod earhting is slightly similar to the pipe earthing. Comparable to pipe earthing, pole earthing requires the hiding of a pole constructed from copper or galvanized iron. Electrodes are installed in the dirt in addition to for that reason decrease the resistance of the earth as required. It's available in both 8-foot as well as 10-foot sizes, with 8-foot being the most common size made use of in household installations.

2.4. Earthing used by Cable.

For cable earthing, a variety of straight trenches are dug. In many cases, rounded conductors are in addition used in the ground. Cable earthing shall be connected with the multi joints as grids. This earthing is mostly used in the High Voltage power substation

3. Variables Impacting Earthing Installations

Several elements can play a role in the earthing installations. These variables will require to be considered for evaluations made concerning the sort of earthing, the type of circuits required, and so on.

The kind of soil resistance is crucial for recognizing the effectiveness of the earthing. The earth's resistance, soil moisture level, salts in the soil, etc. will play a significant function in developing the earthing.

In addition to the soil, the location of the earth pit is vital to find out exactly how the arrangement requires to be done. They will certainly impact the installments if there are below ground blockages in the type of rock beds.

3.1. Procedure to reduce earth resistance

- Get rid of Oxidation on joints and joints ought to be tightened up.
- Put enough water in planet electrode.
- Used larger size of Electrode.
- Electrodes ought to be connected in parallel.
- Pit of more deepness & size- breadth should be made.

4. What are the recommended values for the Good Earthing

The objective in ground resistance is to achieve the lowest ground resistance value feasible, that makes good sense economically as well as literally, when calling the planet, likewise referred to as the soil/ground pole interface.

Generally, a ground must be "0" ohms of resistance, but it's not one typical ground resistance limit recognized by all certifying agencies.

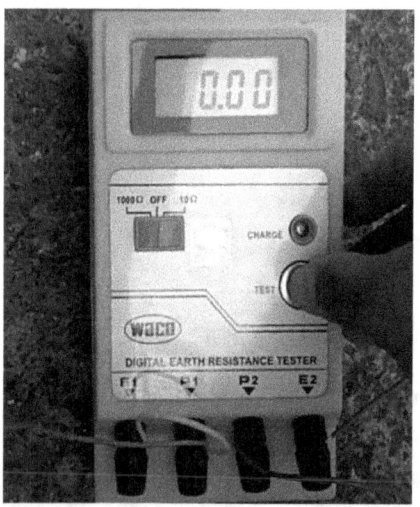

The image shown, Earth resistance tester for Soil and Earth pit resistance test.

Before fix the earth pit, need to confirm the soil resistance, the reason to where can be fix the earth pit by testing the resistance value. The equal distance of 4 temporay rod shall be fixed and connection has to made according to the test kit manual. Press the test button and get the result.

On similar way, 3 equal distance rod metods are using for the measuring points to earth pit resistance.

The NFPA as well as IEEE recommend a ground resistance value of 5 ohms or much less while the NEC has actually specified to "See to it that system insusceptibility to ground is less than 5 ohms defined in NEC 50.56. In facilities with sensitive tools it ought to be 5ohms or much less."

The telecommunications sector has usually made use of 5 ohms or much less as their value for basing as well as bonding while electrical utilities construct their ground systems so that the resistance at a large station will certainly be no more than a few tenths of one ohm.

Page intentionally left

Know about Basic Loads in Electrical

The load which takes energy is called as Electrical load. As we discussed in the previous chapter, the current and transform into light, heat, work and etc. Knowing about this loads types also important where you balance the complete loads. Here, we tried to convey the simple way to explain the loads. However, on the domestic loads are mostly not required to consider these factors widely but on the hand these factors are pretty much important to fix in the system.

The term of technical loads can be classified by as follow,

1. Inductive loads
2. Capacitive loads
3. Resistive loads

The nature of loads depends on the many factors such as demand, power, utilization and diversity factors. Here, we are going to discuss about above loads and with the suitable example from the house.

1. Inductive Loads

The inductive loads have a coil arrangement which can be closed by the current and transform to mechanical energy. Inductive load current waves are lagging behind the voltage waves. The power factor which is maintaining the rated voltage is also lagging.

Inductance measure SI unit is Henrys. Just to remember about this loads are have a two type of power which is real and reactive power. The real power is work which done by the load like motor running. In the other hand, reactive power is based on the magnetic field drawn by the source. Always the Total power will be consumed as combined the both real and reactive power (reactive power is VAR).

Simple understanding of the Current lagging in the inductive loads will discuss after complete the capacitive loads.

2. Capacitive Loads

The principle of capacitive loads is maintaining the VAR (Reactive power) and power factor which cause by lag of inductive loads. It's an inverse concept of the inductive loads. Usually capacitive loads charge and release the power. The circuit voltage always leads the current therefore it can be balance the factors.

Anyhow, these loads are lightly loaded power source where apply to decrease the harmonic currents.

Capacitive charges can be accrued in underground cables, filter circuits in electronics, load banks, Motor startup circuit in the single phase motors, IT company and manufacturing units to lead the power factor as well as allowing reactive power to be allowed from these circuits.

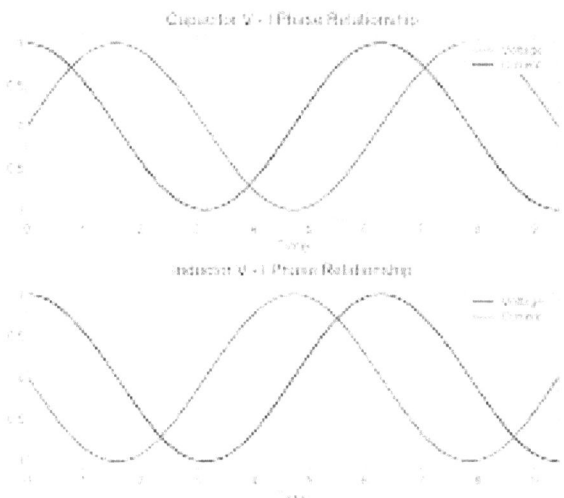

Capacitive loads are results reactive power improves, which can be balance the system voltage to long end user in the line.

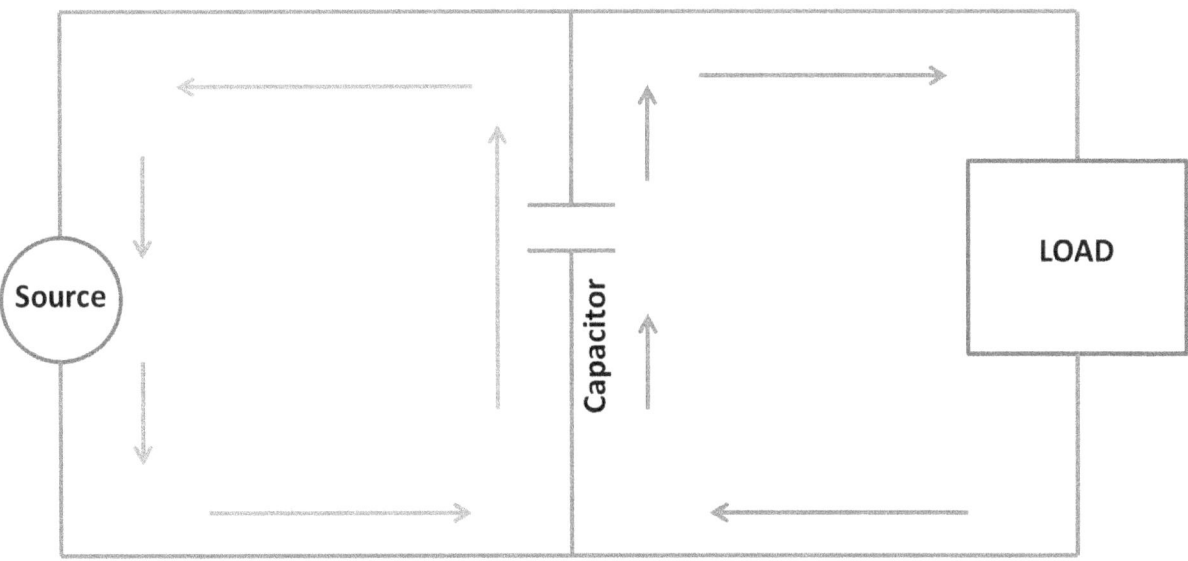

-Reactive Current on full load   -Reactive Current on No load

Just for an understanding about the capacitive loads we will see a small example here, On the power system we have added a capacitor bank 35kVAR to maintain a leading power factor because here we have inductive loads are connected in the power system.

Let say, when the motors running then Capacitor bank will be on to increase the reactive load and result, lead power factor which can maintain up to 1.0.

Instance on the NO load, Capacitor bank connected the result this reverse current will generate the reactive power which can lead to trip the power system under the reactive-reverse power protection.

3. Resistive Loads

The resistive loads are generally involved to current conversion in the form of energy as heat. Main theme is load will not generate magnetic fields like inductive loads. Simple examples are heater and incandescent lights.

The load which can resist the flow of electricity, in doing to convert the luminous or heat dissipated.

BEST EXMAPLE OF RESISTIVE LOAD IS INCANDESCENT BULBS

On the resistive load, Current rises immediately to its standard state value when the current in phase with the enough voltage. An incandescent bulb generates the light by flowing current through the filament by a vacuum form. This phenomenon causes it to heat raise and create the light energy. Heaters are similar to the light but the coil thickness will be mostly hard to just convert as heat dissipation,

HEATERS ARE OTHER BEST EXMAPLE OF RESISTIVE LOADS

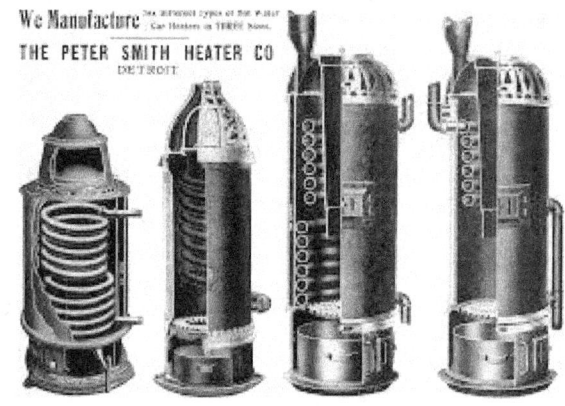

In this type of loads will not take starting current as like an inductive loads. Therefore, resistive loads are little inrush current circuits.

The common resistance SI unit is Ohm (Ω).

Since, It's linear circuit. Voltage and Current in a resistive load are said to be "in phase" to each other. When current rises or falls, parallelly voltage also gets rises and falls with it.

3.1 Optimum application of resistive loads

Voltage optimizations possible in the resistive loads, as resistive loads are made to optimally convert current into power at recommended rated voltages. In order to consumption of power and extend the life of the appliance circuits.

Voltage optimization consider a minimum operating voltage and ensure the power quality supply in order to control the effects of harmful brownouts (Incoming supply voltage drop) and surges (Power spikes which can generate from the source or other faulty loads), where in the light bulbs and heater loads.

So these kinds of circuits are considered as less complex conventional resistive loads and less power consumption. Voltage optimization provides optimal operating extended life and stable power supply.

Page intentionally left

Right Cable Selection and Applications

The most crucial facets of developing any kind of component and also designing of a house electric system are establishing the correct size and type of cable to use for each and every circuit. Also, small cable size will risk creating warmth in the cable; as well as too big size will be throwing away money on copper you do not need.

Additionally, what type should you make use of - plain copper or tinned, typical PVC insulation or thin-wall insulation? The following article must give you an understanding right into just how an electrical cable is defined and permit to you pick the appropriate one for your application

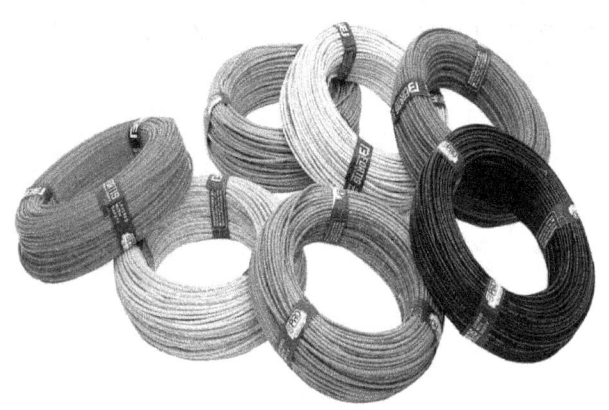

Additionally, what type should you make use of - plain copper or tinned, typical PVC insulation or thin-wall insulation? The following article must give you an understanding right into just how an electrical cable is defined and permit to you pick the appropriate one for your application.

Additionally, what type should you make use of - plain copper or tinned, typical PVC insulation or thin-wall insulation? The following article must give you an understanding right into just how an electrical cable is defined and permit to you pick the appropriate one for your application.

It is important to understand cable construction, characteristics and also rating to comprehend problems connected to cable systems. However, to properly select a cable system as well as assure its adequate operation, added expertise is called for. This understanding might include service conditions, kind of load served, modus operation and maintenance.

1. Cable Spec Requirements

Cables are generally defined utilizing the complying with properties:

1.1. The conductor cross section and complete diameter

Revealed in mm ² as well as explains the overall cross-sectional area of the copper conductor. You will certainly often see cable referred to as 1mm or 2mm cable without the mm² indication but it is very important to keep in mind that this does not mean the diameter of the cord. This can often cause confusion so simply remember that the main spec for wire will certainly be its conductor's cross-sectional area and also cord will certainly never be described by its diameter alone.

Complete over all diameter which is include the insulation and sizing in square mm.

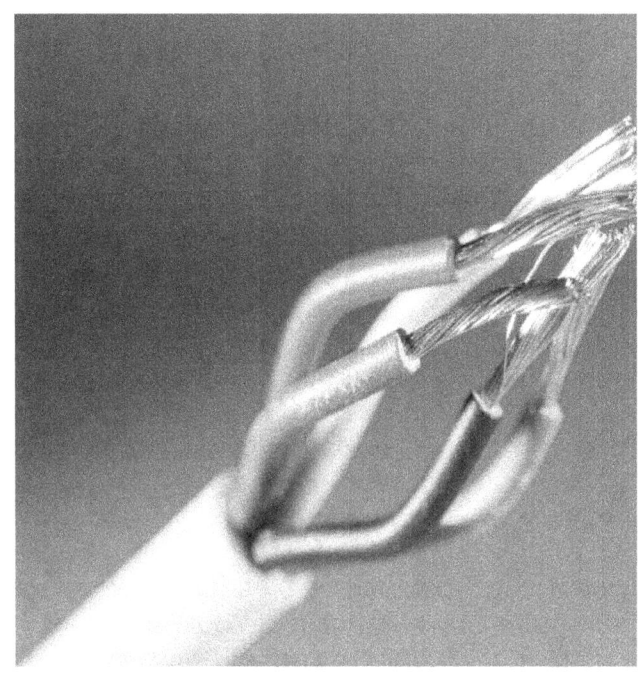

1.2. Size of the cable and Number of the stands in the conductors

Revealed as per number of stands in the conductors and given size. If, 28/0.30 means, 28 stands in the conductor and each of the stands in 0.30 diameter.

1.3. Rated current rating

It's mentioned in the ampere (A) as we discussed in the previous chapter, and what is the maximum current can carry continues in the safe manner.

1.4. Resistance of the Cable

Another important factor is resistance of the cable and it's expressed in Ohms per meter and this can help to finding the voltage drop in the line. In order to this some cable which can resist the more heat or chemical effect for special applications.

1.5. Selection of the Cable

On this chapter we will check some major points which are considered before decide the cable.

2. Capacity of carrying the Load

Each component or appliance linked to a circuit will certainly have a current draw connected with its duty may or may not continuous and also it is essential that the providing power to these can carrying the typically expected current, with a safety margin . After that it is most likely to result, the wire ending up being warm and potentially capturing fire, if it is not qualified or proper selection.

Also fuses in the circuit to safeguard the wire, the cable itself ought to be of an adequate rating to avoid this over-heating taking place under typical situations.

As we already discussed the basic of current, since P=V I

I= P/V, where the simple example below,

If, we wished to wire up a blender that we know has a power ranking of 700W, after that utilizing I = P/V the current draw would be 700W/230V = 3.04 A. This informs you that you can utilize a cable with a rating of 3.04 A or above, nevertheless it is good method not to select a circuit operating at the upper end of the cable's rating and so you should choose a cable with some additional capability. In this situation 0.75 mm ² (11A) wire would certainly be appropriate.

3. Voltage drop in the Cable

All components of an electric circuit have resistance, including electric cable, which indicates that there will certainly be energy loss in the form of potential drop happened along the size of the cable. Equally as a bulb converts electrical energy right into warm as well as light due to its resistance, and so causes a potential drop, a copper conductor has resistance and also will certainly convert some of the power it conducts, creating a voltage drop in the same way. The difference is that voltage decrease throughout a light bulb (or various other tons) works as that's what makes it work, however voltage decrease along cable and other passive parts of a circuit is not desirable as it's not an useful conversion of energy.

In low voltage systems cable length can have a considerable effect on potential drop. Even a cable run of a couple of meters for small cross-section conductors can generate significant voltage drop as well as this issue is shown well on some appliance where the lights or music system circuits are not as brilliant as they could be. If you inspect the voltage at the sockets you might discover that system not getting a complete voltage from the circuit because of the conductor dimension being as well less for the wire run length.

So we want to select a cable to see to it that the voltage drop is not so large that it will cause problems, but what serves as well as exactly how do we determine the right cable dimension to utilize? Well the generally acceptable voltage drop for DC/AC circuits is around 3-4% as well as we can use V = IR to compute the voltage drop for a cable if we know the current draw of the circuit and the cable's resistance per meter.

For an example of Voltage drop

Using the above instance of a 700W blender we know it takes 3.04 A, so if we were to utilize a 0.75 mm² wire which has a resistance of 0.026 W/m and its total length from source to device is 3meters, then the voltage drop would certainly be:

V. Drop = IR = 3.04 A x (3m x 0.026 W/m) = 0.237V or 0.10% on 230V.

This shows that although 0.75 mm² cord is good for the anticipated current draw of the light, it's not OK for the wire run size as the decrease is higher than 3%.

So what concerns 1.0 mm² wires with a resistance of 0.018 W/m?

V. Drop = IR = 3.04A x (10m x 0.018 W/m) = 0.164 V or 0.07%.on 230V

This shows that 1.0 mm² wire (at a present rating of 16A) will appropriate for the cable run length as the drop is well under 1%.

There is a general rule that states if you're uncertain whether the conductor is big enough for the job, rise a size. This is a bit crude and not very scientific however it's not a bad policy to apply as boosting wire size can't do any damage.

4. Conductor type Selections

The Low Voltage and Telecommunication cables are available in different types and constructions, Choose the material of the Cable is interesting chapter again,

Kinds of Wires- There are generally 5 kinds of cords are widely using in the market:.

Just before enter to this subject; it's required to know about the cord lettering.

The letters THHN, THWN, THW and also XHHN represent the primary insulation kinds of individual cables. These letters depict the complying with NEC requirements:.

- N-- Nylon finish, resistant to damage by oil or gas.
- X-- Synthetic polymer that is flame-resistant.
- T--(Thermoplastic)Polycarbonate insulation
- H-- Warm (heat) resistance
- HH-- High heat resistance (as much as 194 ° F).
- W-- Suitable for wet areas

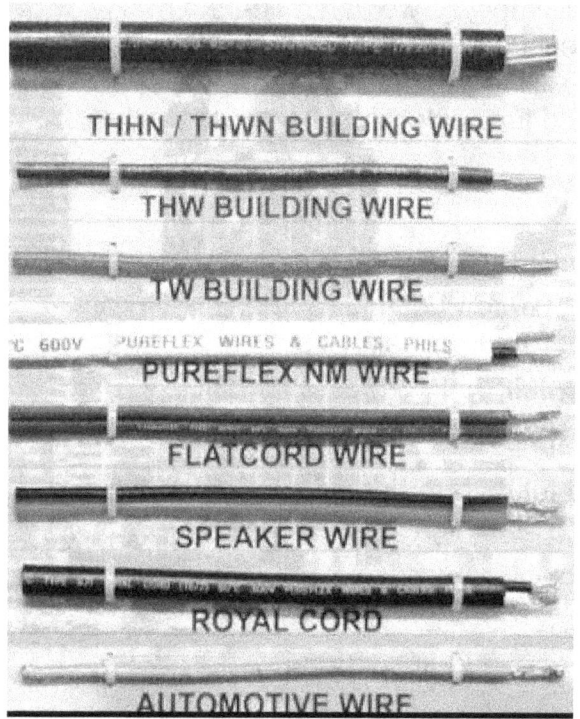

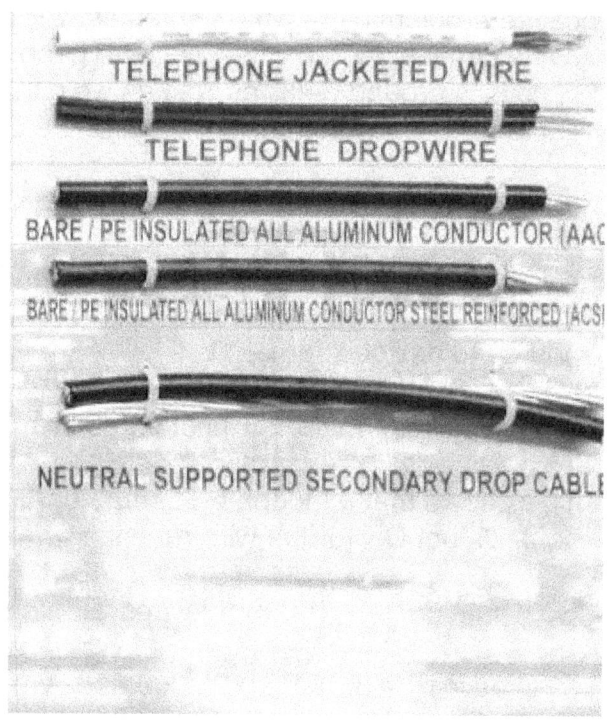

I. **Triplex Wires**: Triplex cables are normally utilized in single-phase solution drop conductors, between the power post and also weather heads. They are made up of two shielded lite weight aluminum wires wrapped with a 3rd bare wire which is made use of as a typical neutral. The neutral is usually of a smaller size and also based at both the electric meter and also the transformer.

II. **Key Feeder Wires**: Main power feeder wires are the cords that connect the Electrical post to main distribution box at the house. They're made with stranded or strong THHN cord and the wire set up is 25% more than the draft load.

III. **Panel Feed Cords**: Panel feed cables are typically black protected THHN cable. These are made use of to power the major junction box as well as the breaker panels. Much like major power feeder wires, the cable ought to be ranked for 25% more than the actual draft loads.

IV. **Single Hair Wires**: Single strand cable additionally utilizes THHN cord, though there are various other versions. Each wire is different as well as numerous wires can be accumulated through a pipeline conveniently. Single strand cords are one of the most prominent options for layouts that make use of pipelines to include wires.

V. **Non-Metallic Sheathed Wires**: Non-metallic sheath cable, or Romex, is made use of in the majority of houses as well as has 2-3 conductors, each with plastic insulation, and also a bare ground wire. The specific cords are covered with an additional layer of non-metallic sheathing. Because it's reasonably cheaper and also readily available in rankings for 15, 20 and also 20 amps, this kind is liked for internal electrical wiring.

5. General Color codes of Cable

There are several cable recognition standards, and a lot of them rely on shade codes. Not all cord shade codes coincide, though, and some even oppose each other. Which typical should be used in your facility? It depends on your place, installation type, voltage, and various other factors.

Keep in mind that older installments might use various color codes. In workplaces, it's an excellent concept to record the shade code that is being complied with. This way, job will be more secure, and also future upkeep will certainly be much easier.

Page intentionally left

6. UNITED STATE Wire Shade Codes

In the USA, the complying with color codes are typically used for power cables in "branch circuits," the circuitry in between the last safety device (such as a breaker) and also the circuits (such as a device or home appliance).

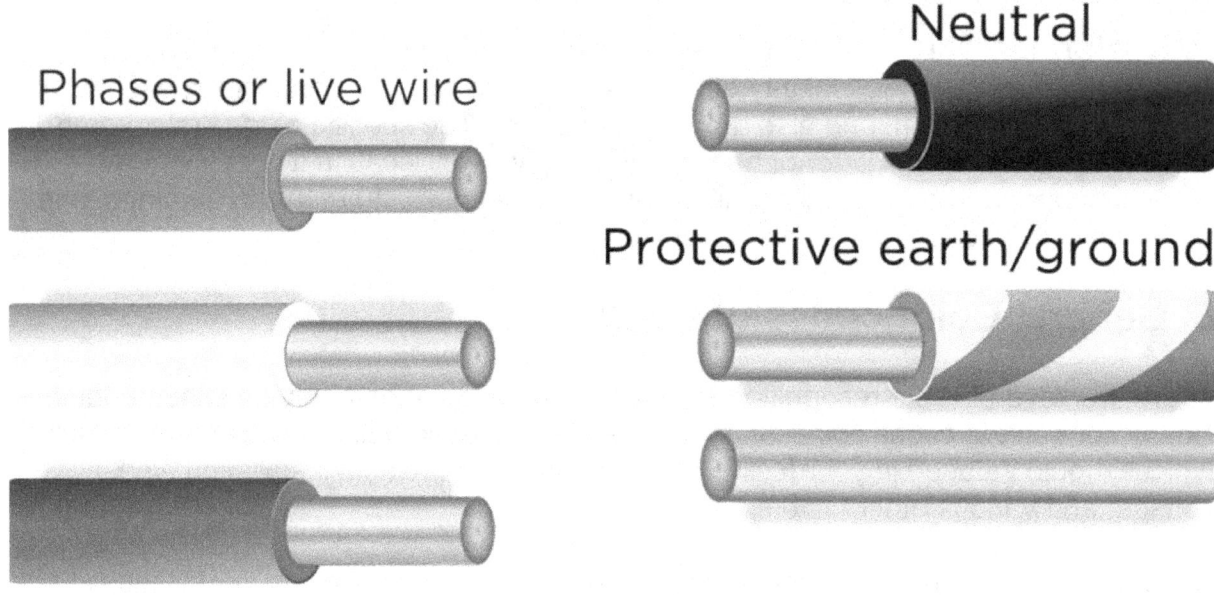

120/208/240 Volt Alternative current Cable Colors

These systems are common in residence and also workplace environments.

- Phase 1 - Black
- Stage 2 - Red
- Phase 3 - Blue
- Neutral - White
- Ground – Yellow, Green and yellow shade

If the wiring system has one phase at a higher voltage than the others, making use of a "high-leg" link, that phase's wires ought to be noted with orange. (This is needed in NEC short article 110.15.) However, these high-leg delta systems are unusual with newer installations.

There are numerous cable identification criteria, and several of them rely on shade codes. Not all cord color codes are the very same, though, as well as some even negate each other. If the electrical wiring system has one stage at a greater voltage than the others, utilizing a "high-leg" link, that stage's wires need to be marked with orange.

Importance of Light design and Construction

Knowing about the light design and construction are very much important in the electrical system. In this world no lives are there without light energy from sun and moon, secondly electrical lights. Generally, Lights are consuming up to 20% of the typical family electricity spending but plan to concerning 6% of its energy use with help of installed illumination modern technologies, lighting design as well as individual practices can make a distinction. Properly designed and also effective illumination can produce household power cost savings.

Always the goal to light your house

1. Give a risk-free, comfy, aesthetically appealing and desirable setting
2. Be as energy efficient as possible.

To make more interesting on this topic, we are mentioned a topics here and technique is using nature light and electricity light to achieve an efficient and energy saving.

1. **Using sun (Day lights)lights to room**
2. **Electric lighting style elements**
3. **Varieties of lighting requirements as per their usage**
4. **Key points angle for directional lights**
5. **luminous flux**
6. **Some standard principles**
7. **Leading 10 actions to illumination**

Fix the light-design before search in market

Thoughtful lighting style combines numerous daylighting as well as electric lighting strategies to optimize the distribution of light inside the building. It thinks about entire structure power effects to minimize the building's overall power usage as well as integrates the design of daylight access (with skylights and windows) with electrical lighting, including controls. It benefits from shading techniques as well as polishing modern technologies to moderate the intensity and spectrum of the daylight confessed to the house, to minimize warm gain throughout the cooling period as well as heat loss during the home heating period. It selects the best window aperture sizes, glazing and also shielding layout for each positioning to reflect the expected solar angles, warm gain and glare requirements (see Easy layout; Style for environment; Alignment; Shading; Glazing; Skylights).

Effective lights design indicates placing light where it's desired and also required, as well as minimizing or getting rid of light in other places.

1. Daylighting design facets

The scientific research of 'daylighting' deliberately uses daytime to lower or negate the requirement for electric light. Sources of daytime consist of sunlight, which is an extremely bright, directional light beam, and skylight, a scattered light of regarding one-tenth the illumination of sunlight. Daylight is dynamic, continuously changing its features (intensity, colour, direction).

Style your new home to not need electrical lights throughout daylight hours.

A goal of all brand-new houses must be to not need any electrical lights during daylight hours. Siting, positioning and also size of the home entered into play but every factor to consider should be provided to reducing reliance on electric lights throughout daytime hours.

Done appropriately, daylighting layout can provide saving on energy consumed by the structure. Done inaccurately, it most generally boosts the warm load on the home and its cooling power consumption. If the daytime control system is inadequately implemented, building residents need to deal with glow and/or thermal discomfort making use of one of the most practical methods available (e.g. curtains attracted, running ac system), which subsequently negates any advantage that daylighting might have used (see Passive style; Shading).

1.1. Simple principles for daylighting

➢ Skylights and light tubes of appropriate sizing and also layout can allow light without adding warm in summertime or shedding warmth in wintertime.
➢ Externally reflected daylight consists of much less heat than straight permeating sunlight (i.e. the infrared heat is primarily soaked up by all-natural and also developed environments).
➢ Light coloured interior surface areas show much lighter and also lower the degree of artificial lights needed.
➢ Clerestories (with the associated eaves suitably sized) are really efficient at supplying daytime to the core locations of a residence.
➢ Bright areas can manipulate tubular daylighting gadgets-- tubular skylights-- which send out direct-beam sunlight into the room listed below as well as can supplying very high illumination levels when the skies is clear.
➢ Direct sun must be left out from task areas (specifically sleek surface areas including kitchen area benches and desktops) due to the high capacity for glare and pain.
➢ Interior sunlight infiltration can be managed with the least effect on an outside sight by vertical blinds on mainly east as well as west oriented windows as well as straight blinds for predominantly north (and southerly, for north of the tropic of Capricorn) oriented windows.

1.2. Easy daylighting suggestions

To light task areas efficiently, straight natural lights from windows, skylights or light tubes (see Skylights) need to be close to the job location. Windows must become part of the border wall of the area being lit and also skylights need to be located in the roofing directly over the job location to be lit.

Thoughtful lighting style incorporates several daylighting and electrical lighting approaches to optimize the distribution of light inside the building. It takes into consideration entire structure power impacts to minimize the structure's general power use and also integrates the style of daytime access (with windows and also skylights) with electric lighting, including controls. The scientific research of 'daylighting' deliberately utilizes daytime to minimize or negate the requirement for electric light. Resources of daytime include sunshine, which is an extremely bright, directional beam of light, and also skylight, a scattered light of concerning one-tenth the lighting of sunlight. If the daylight control system is inadequately applied, building owners have to deal with glow and/or thermal discomfort utilizing the most suitable ways at hand (e.g. drapes drawn, operating air conditioner), which in turn negates any advantage that daylighting could have provided (see Easy style; Shading).

Light tubes can be purchased for just in less amount and also set up conveniently by a tradesperson or capable DIY proprietor. They can change a 60W light competing approximately 8 hrs a day in a poorly lit space, saving as high as $30 each year. They can thus spend for themselves in less than 5 years (varies according to size of use and also consolidated power level of lights on button circuit).

Light shelves mirror daytime to penetrate deep right into a structure. They are suitable for north as well as south elevations however not the level sun angles of eastern as well as west. A straight overhang with a high reflectance upper surface is positioned above eye-level to mirror daytime onto a light coloured ceiling and also much deeper into a room.

By varying the height, angle as well as outside or interior estimate of a light rack, you can control the pattern, intensity and also deepness of infiltration of all-natural light (including sunlight) within a space. Light racks need to be light in colour and also call for regular cleansing.

Not only do light shelves permit light to pass through much deeper into the area, they can shade near the windows to minimize window glow or produce a sunlight patch. Exterior shelves usually give much more efficient color while interior shelves supply deeper showed light. A combination of outside and also interior racks works best to provide an also illumination gradient.

Glass blocks are a helpful resource of daylight in walls that are close to limits or require personal privacy. Glass brick panels allow diffuse daytime while keeping noise as well as aesthetic personal privacy, as well as fire ratings.

Glass bricks allow diffuse daytime while keeping audio and also visual personal privacy in wall surfaces that are close to borders or encounter the street.

2. Electric lighting style elements

Use of electrical lighting in the room has two aspects: particular job lighting and also developing a day-time environment for a space or area.

Human vision has a really high vibrant variety however understanding of illumination changes with the total brightness of the entire area. The eyes adapt to reduced light levels at evening as well as it is unnecessary to attempt to replicate the high degree of illumination available from daytime.

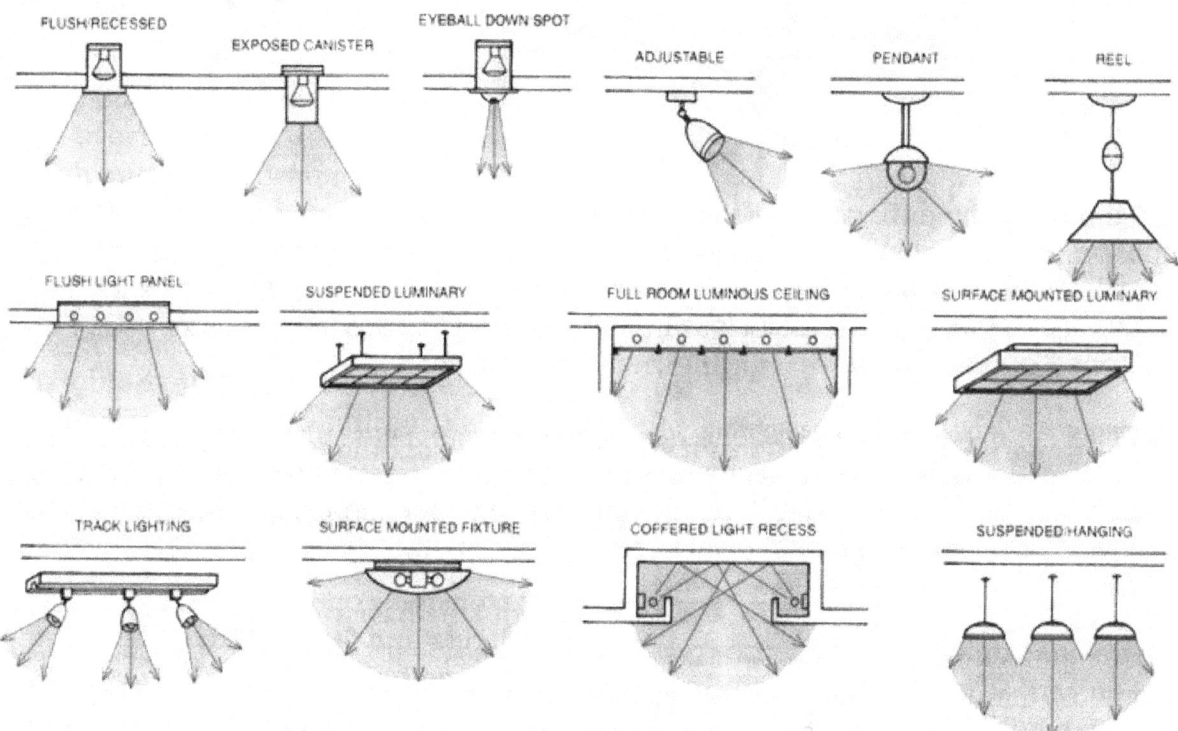

When considering lighting an area, service sights within. The human eye is drawn in to intense things as well as necessary must be awarded with something of rate of interest. By contrast, dark areas are of limited destination but offer to highlight (by comparison) the brighter objects of passion. Usage highlights (concerning 10 times the ambient light level) to draw attention to crucial objects or spaces in an area, or for illumination particular jobs. Very carefully choose features to highlight (e.g. art work, sculptures, and also furnishings items) and also utilize the minimal efficient emphasize level so you do not throw away energy.

Reading lamps or table lights are an efficient, effective and also flexible ways of offering greater task lighting instead of boosting general lighting of the whole area. They can also become part of the accent illumination for state of mind setting (e.g. table lamp on side table in lounge room).

2.2. Applications for electric lights

Consider specifically the areas that serve greater than one function and require greater than one style of lights (e.g. media watching, studying, basic activity). Usage separate illumination solutions as well as circuits for each and every feature as opposed to integrating them right into a single circuit. Lights might require being on separate switches, and/or dimmers used to create the lighting preferred.

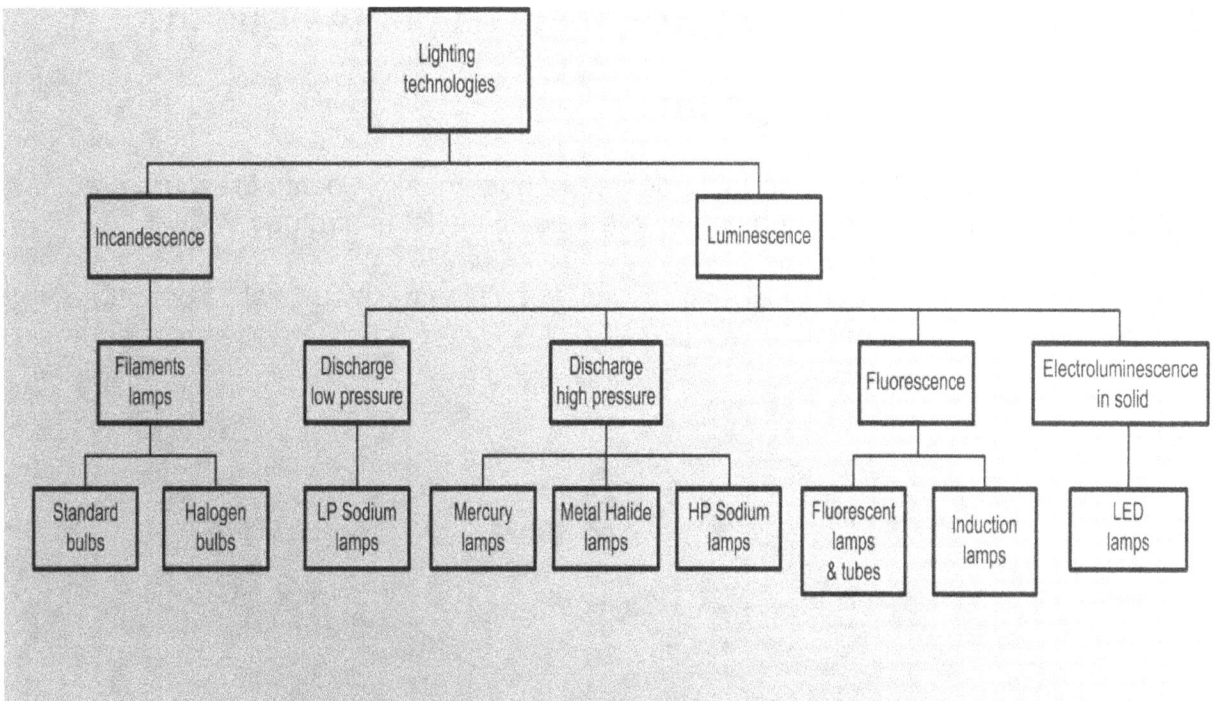

Each light has advantages and also drawback and excellent style utilizes a suitable kind for every application.

Remember:

1) Light is heavily soaked up (wasted) in dark-coloured spaces.

2) Light can be indirectly mirrored (i.e. cove as well as pelmet lighting) to produce very subtle background lighting however just in light-coloured rooms/surfaces.

There is no 'ideal' lamp for all applications. Of the many alternatives, each has advantages as well as downsides: great style uses an ideal lamp/light fixture for each application.

The numerous light innovations generate light in different ways. Choose lamps ideal fit to creating wanted lighting impacts such as light circulation, switch-on time and dimmability.

For instance, some small fluorescent lamps (CFL) take a few secs to strike as well as 'warm up', and are hence unsuitable where use might just be for a few secs (e.g. cooking area pantry) or where lights are switched on as well as off swiftly. CFL lamps, although efficient, are an inferior choice in these rooms to tungsten halogen or perhaps LEDs, although LEDs may not have the expense advantage for such brief uses.

Many spaces need two kinds of lighting: general lights and also task/accent lights. Usage different lamps and light installations for each objective.

3. Varieties of lighting requirements as per their usage

3.1. General/ambient lighting

General lights that emits a comfortable degree of brightness. A central resource of ambient light in all rooms is fundamental to an excellent lights strategy.

a) Use omni-directional (light in all instructions) lamps in necklaces, light fixtures, ceiling or wall-mounted components.

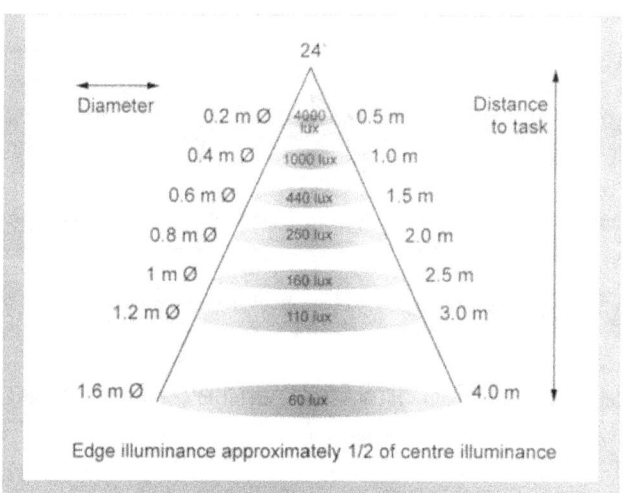

b) Avoid making use of downlights for basic illumination. They make bright 'pools' of light on the flooring (most flooring surface areas take in as long as 80% of the light) while making the ceiling cavity show up dark, which creates a 'dismal' feel. Downlights are much better suited to task illumination over work spaces. Approximately 6 downlights can be required to light the same area as one necklace light. Consider various other means of lights with fluorescent omni-directional lights prior to mounting downlights or if used, fit reduced power level and extra effective bulbs.

c) Choose light installations and also lamp shades that enable the majority of the light through so a reduced power level lamp can be utilized. Some light installations can block or take in 50% or more of light.

3.2. Task/accent lighting

Task lights is made use of to illuminate specific jobs such as analysis, embroidery, cooking, research, leisure activities or games. Accent lighting adds drama to a space by developing visual passion. It can stress paints, residence plants and collectables, or highlight the structure of a wall surface, drape or outside landscape design.

- Directional lamps or downlights, such as LED or halogen reflector lamps, are best employed for this objective.
- Use desk/table/floor lights in areas where the activity or furniture is most likely to turn (lounge, eating, rooms).
- Where lit up task surface areas will not alter (e.g. over cooking area benches), usage fixed directional lights.
- Make certain job lights is devoid of distracting glow and darkness however intense sufficient to avoid eye stress.

4. Key points for choosing the appropriate light beam angle for directional lights:.

Basic rule: for the exact same power level light, the smaller sized the beam of light angle the brighter the surface brightened but the smaller the area brightened.

Select the suitable beam of light angle by determining the biggest dimension of the feature to be lit and the distance from it. The product packaging of many directional lights usually reveals a simple graphic to aid select the appropriate light beam angle.

They make bright 'pools' of light on the floor (most floor surfaces soak up as much as 80% of the light) while making the ceiling tooth cavity show up dark, which produces a 'dismal' feel. Up to 6 downlights can be needed to light the exact same area as one necklace light. Believe regarding various other methods of illumination with fluorescent Omni-directional lights prior to mounting downlights or if used, fit lower power level as well as more reliable bulbs.

Task illumination is used to light up certain tasks such as analysis, stitching, food preparation, research, pastimes or games.

5. Luminous Flux (Choosing lamps)

5.1. Colour of light.

2 things are used to define the colour of white source of lights:.

a. Colour rendering index (CRI).
b. Correlated colour temperature (CCT).

Colour designation	CCT (K)	Appearance	Typical uses
Warm white	2,700-3,200	Similar to incandescent	Household rooms
Cool white	4100	Neutral light	Offices, garages, workshops
Daylight	5,500-6,500	Cold, harsh, unrelaxed light	Bathrooms, laundries

CCT, determined on the Kelvin (K) temperature range, explains the 'color' of white light emitted.

The product and colour of your furniture can contribute in your choice to make use of cozy or trendy lights, since the variant of lighting colour can make room colours appear quite boring or really vivid.

5.2. Correlated colour temperature.

Unit: Kelvin.

Role: Range to describe exactly how 'warm' or 'awesome' the light appears.

Origin: In theory, as a things (e.g. piece of metal) is heated up, it shines, changing colour from a red to orange to yellow to white to bluish-white as the temperature boosts.

5.3. CCT of regular domestic lights:.

Incandescent lamps: operate by heating up the filament to 2,700 K and necessarily, have a colour temperature of 2,700 K.

Fluorescent, CFL as well as LED lights: offered in a wide range of colour temperatures.

Awesome white (left) and also warm white (appropriate) colour temperature level lamps provide areas a various appearance.

Match the lamp's colour temperature to the tones of your space. Warm colour temperatures make warm colours like reds, browns and yellows well; cool colour temperature levels render great colours like greys, eco-friendlies and blues much better. In areas adhering to these basic rules, furnishings appear even more vibrant. If you have a mix of home furnishings, use lamps that produce light in around the 3,500 K variety.

No matter what colour temperature level light you select, if it has a low colour making index, nothing will look great under it.

5.4. Colour rendering index.

Unit: None.

Role: Range in between 100 and listed below 0 where 100 represents true natural colour recreation for that particular colour temperature level.

Beginning: A reference source such as sunlight is specified as having a CRI of 100; incandescent lights radiate a similar range of light to the sun.

5.5. CCT of typical domestic lights:.

Incandescent lights: 100 Fluorescent, CFL lights: 60-95 LED lamps: 80-90.

The colour making index rates the portrayal of colour.

CRI ranks the capability of the light to accurately depict colours of objects in the room being lit.

A CRI of higher than 80 is usually sufficient however, for specialised tasks where colour is necessary (cooking, using makeup, painting) it is recommended to select lights with a CRI over 90.

Lights of the same colour temperature can vary in their capability to make colours correctly.

6. Some standard principles:.

One button to transform on all lights in a huge room is really inefficient. Place switches over at departures from spaces and use two-way changing (for lengthy corridors or stairwells) to urge lights to be transformed off when leaving the room.

a) ' Smart' light buttons and also fittings use movement sensing units to transform lights on and off instantly. These are useful in areas used rarely where lights may be left on (for long times) inadvertently, or for children, the elderly and individuals with specials needs. Built-in daytime sensing units make certain the light doesn't switch on unnecessarily during daylight hours.

b) Use timers, daytime controls as well as activity sensors to change outdoor safety lights on and off instantly. Similar controls are particularly useful for usual locations, such as corridors, stairwells as well as passages, in multi-unit real estate. Some controls are not compatible with certain light types so consult.

c) Think about utilizing solar energy lights for garden and sensing unit safety and security lights.

d) Modern dimmer controls conserve power as well as additionally enhance light life. Nevertheless, lowering light outcome to 50% conserves just around 25% of the power (for a halogen light). If you dim some lights a lot of the time, think about changing them with lower wattage lights.

e) Many standard fluorescent as well as LED lights can not be dimmed (although this is boosting), but special dimmers and lamps are available (check packaging or manufacturer's site for details). When installing brand-new light fittings and also controls, check on compatibility.

f) Cozy colour temperature levels render warm colours like browns, yellows and also reds well; cool colour temperature levels make trendy colours like greys, eco-friendlies as well as blues much better. If you have a mix of furnishings, use lamps that create light in roughly the 3,500 K range.

g) ' Smart' light switches and also fittings use motion sensing units to transform lights on and also off automatically. Decreasing light result to 50% conserves just about 25% of the energy (for a halogen light). If you lower some lights many of the time, consider replacing them with lower wattage lamps.

h) Incandescent, CFL as well as straight fluorescent lamps are regulated for energy effectiveness as well as light quality. Various other light modern technologies can differ substantially in quality. Read packaging details as well as technical requirements thoroughly to ensure the product appropriates for your intended use.

Page intentionally left

Bulb Type	Bulb Image	Quantity of light (lumens)	Lifetime (hours)	Colour rendering	Dimmable	Lifetime (hours)	Lamp Colour Range	Efficiency (lumens/W)
LED		med	15-50k	excellent	many dim	15-50k	7000K 5700K 4000K 3500K 3000K 2700K	20-40
CFL		high	5-20k	excellent	few dim	5-20k	7000K 5700K 4000K 3500K 3000K 2700K	> 40
CCFL		med	5-20k	excellent	many dim	5-20k	7000K 5700K 4000K 3500K 3000K 2700K	20-40
Linear & circular fluoro		very high	> 20k	excellent	few dim	> 20k	7000K 5700K 4000K 3500K 3000K 2700K	> 40
Induction		high	> 20k	good	few dim	> 20k	7000K 5700K 4000K 3500K 3000K 2700K	> 40
Halogen		high	< 5k	excellent	all dim	< 5k	2700K	< 20

7. Leading 10 actions to illumination

1) Style a home to not need lights switched on during daytime hrs.

2) Think about the positioning and also layout of areas to ideal usage readily available daytime.

3) Usage surface area reflectance of light coloured surface areas, as well as well positioned pendant and wall lights, permanently light circulation in a room.

4) Choose the type or kinds (e.g. basic illumination, mood/background lights, job lights) of atmosphere you want to create in each room during night-time use.

5) For greater than one sort of atmosphere, readjust light degrees (dark lights) or transform different lights on or off with different changing circuits.

6) Create task or accent illumination with directional lighting.

7) Create general lighting with non-directional lighting.

8) Use warm coloured lights for the house, other than perhaps for restrooms as well as washings where the cooler coloured lights present a look of a clean, sterilized space.

9) Before choosing a lamp, identify relevant characteristics for enlightening each space (e.g. quick startup, lengthy life lamp, dimmable, multi-way changing).

10) For getting the 'correct amount of light' to develop the ambience you desire, think about lumens, which measure of the total quantity of noticeable light produced by a source, not power level (power).

Page intentionally left

Know about Wiring systems and Benefits

Electric Electrical wiring is a process of linking wires and cables to the related gadgets such as fuse, changes, outlets, lights, fans etc to the main distribution board is a specific structure to the utility pole for proceeds power supply.

1. Approaches of Electric Wiring Equipments

Electrical wiring (a procedure of attaching numerous devices for circulation of electrical power from source's meter board to home devices such as lamps, fans and various other domestic devices is known as Electrical wiring) can be done making use of 2 techniques which are.

- Looping wire system
- Joint box or Tee system

1.1. Looping System.

This approach of circuit is globally made use of in wiring. Lamps as well as various other appliances are linked in parallel so that each of the appliances can be controlled individually. All connections at a light or button, the feed conductor is looping in by bringing it directly to the terminal and after that lugging it onward again to the following indicate be fed. The switch and light feeds are lugged round the circuit in a series of loops from one point to another until the last on the circuit is reached. The phase or line conductors are knotted either in switchboard or box and also neutrals are looped either in switchboard or from light or fan. Line or stage must never ever be looped from light or fan.

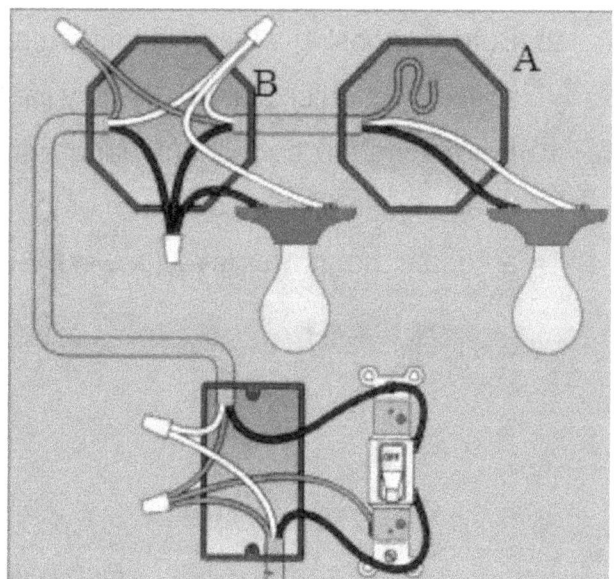

1.1.1. Benefits of Loop-In Technique of Circuit.

1. It doesn't need joint boxes therefore cash is conserved.
2. In loophole-- in systems, no joint is concealed underneath floors or in roof spaces.
3. Fault place is made easy as the factors are made only at electrical outlets to ensure that they come.

1.1.2. Drawbacks of Loop-In Technique of Circuit.

1. Length of wire or cords needed is more and also voltage decline as well as copper losses are consequently extra.
2. Looping-- in Switches as well as light holders is normally challenging.

1.2. Joint Box or Tee or Jointing System.

In this technique of circuit, links to home appliances are made with joints. These joints are made in joint boxes using ideal adapters or joints cutouts. This technique of wiring does not eat way too much wires dimension.

You could believe since this technique of electrical wiring doesn't need way too much wire it is as a result less expensive. It is of course however the money you saved from acquiring cables will certainly be utilized in purchasing joint boxes, therefore formula is well balanced. This method is suitable for temporary installations and also it is cheap.

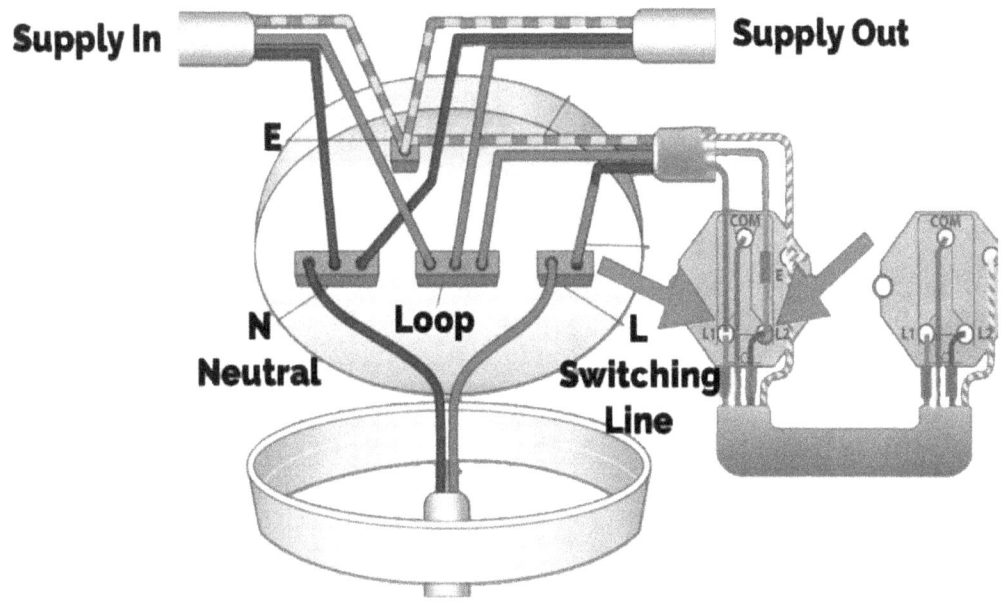

1.3. Various Types of Electric Wiring Systems.

1. Cleat circuits
2. Wood case and capping electrical wiring.
3. Lead sheathed electrical wiring.
4. Conduit wiring.

1.3.1. Cleat Circuit.

This system of electrical wiring consist of average VIR or PVC insulated cables (periodically, sheathed and also weather condition evidence cable connections) braided and also compounded hung on walls or ceilings using porcelain cleats, Plastic or timber.

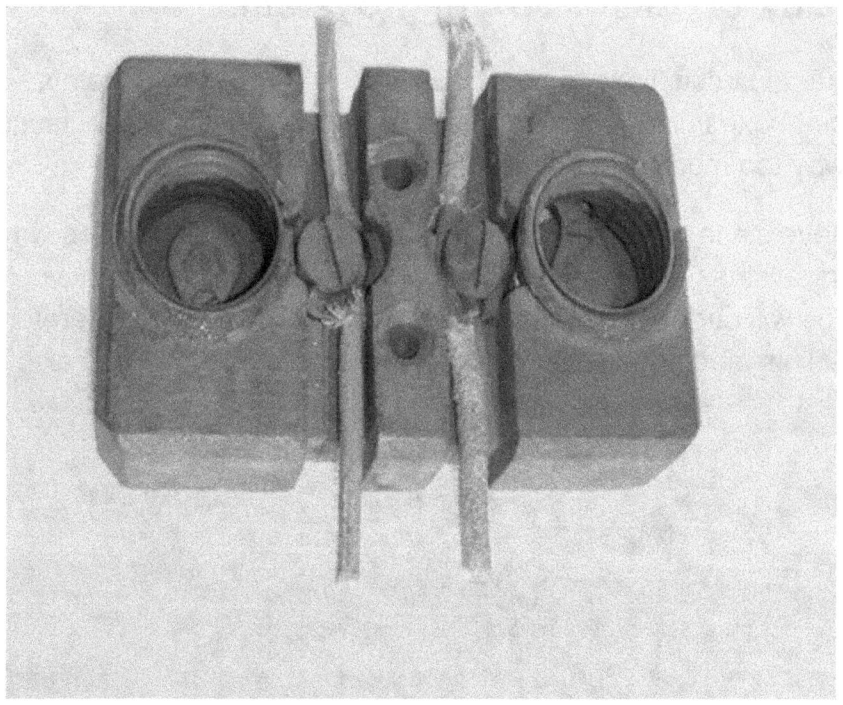

Cleat circuit system is a short-term wiring system for that reason it is not suitable for domestic properties. It's rare to use on nowadays.

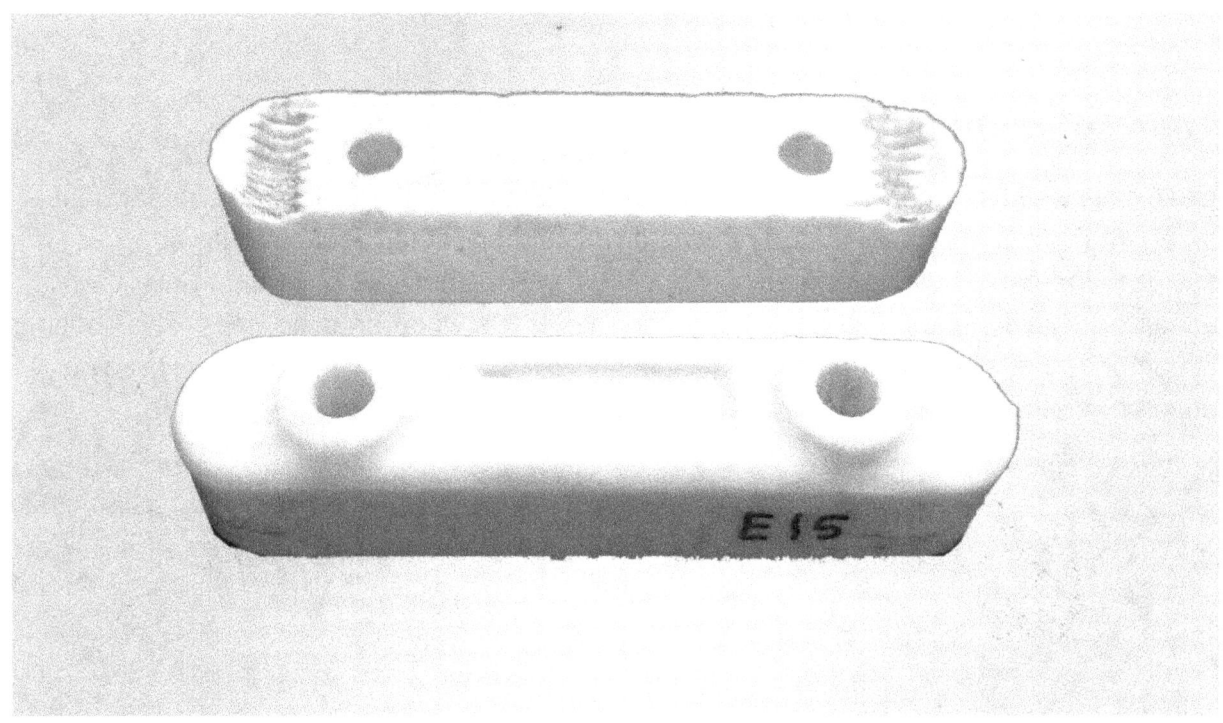

1.3.1.1. Merits of Cleat Wiring:.

1. As the wires as well as cables of cleat wiring system is in outdoors, As a result fault in wires can be seen as well as fixing quickly.
2. Cleat wiring system setup is easy as well as basic.
3. Customization can be conveniently performed in this wiring system e.g. change and addition.
4. Evaluation is simple and easy.
5. It is cheap and simple wiring system.
6. A lot of suitable for momentary usage i.e. incomplete building or military camping.

1.3.1.2. Downsides of Cleat Circuit:.

1. In this circuit system, the cables and also electrical wiring is in open air, as a result, oil, Steam, humidity, smoke, rain, chemical and acidic effect may harm the wires and cables.
2. Look is not so good.
3. Cleat circuit cannot be use for irreversible usage due to the fact that, Droop might be occur after at some time of the use.
4. There is constantly a danger of fire and also electric shock.
5. It can't be used in important and also sensitive place and also places.
6. It is not long-term, sustainable and reliable wiring system.

1.3.2 Capping and casing wiring

Casing as well as Topping wiring system was popular circuit system in the past but, it is taken into consideration outdated now as a result of sheathed and conduit wiring system. The cables utilized in this sort of circuit were either VIR or PVC or any other accepted shielded cable. Here, the carrying medium is wood housing units. The housing is comprised of a strip of timber with parallel grooves reduced size wise so regarding accommodate VIR cords. The grooves were made to separate contrary polarity.

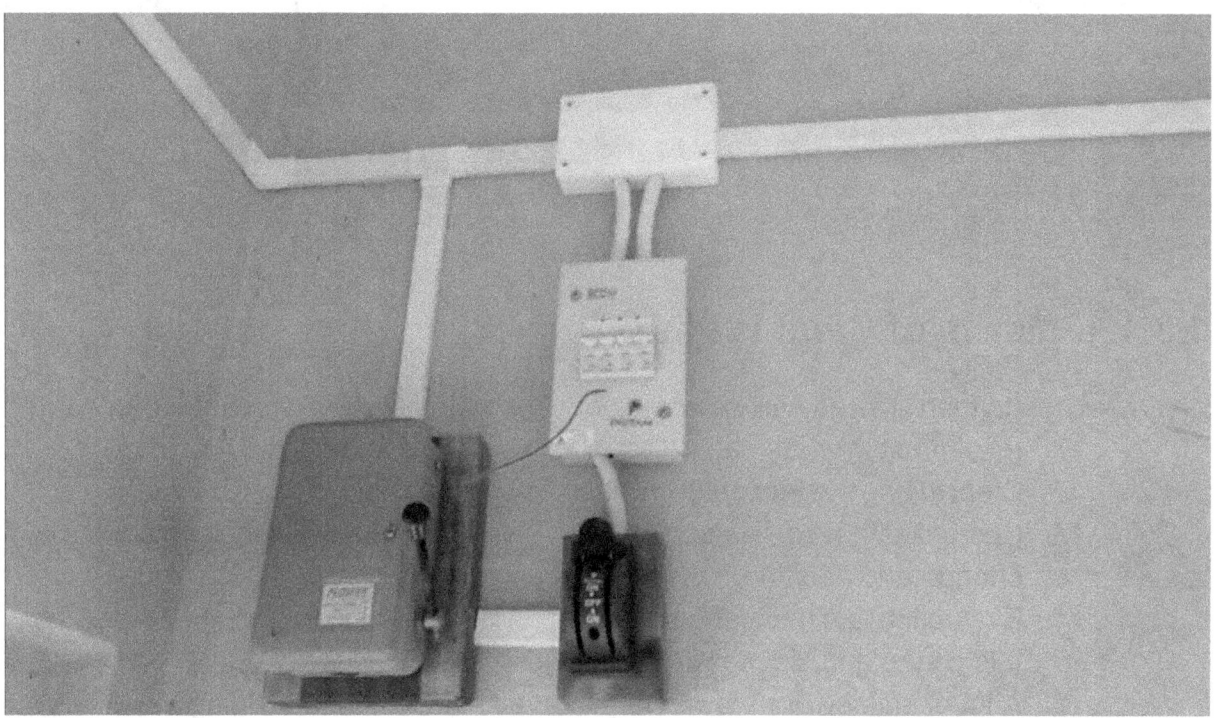

1.3.2.1. Advantages of Covering Capping Electrical Wiring:.

1. If Phase and Neutral cord is mounted in separate ports, after that fixing is easy.
2. Remain for long time in the field because of strong insulation of casing and capping.
3. It stays secure from oil, Rain, smoke and steam.
4. No danger of electrical shock as a result of covered cords as well as wires in case & capping.
5. It is cheap wiring system as compared to sheathed and conduit electrical wiring systems.
6. It is solid and durable wiring system.
7. Personalization can be conveniently performed in this wiring system.

1.3.2.2. Disadvantages Case Capping Circuit:.

1. Costly fixing as well as require even more material.
2. Product can't be discovered quickly in the modern.
3. White ants might harm the housing & capping of timber.
4. There is a high risk of fire in housing & covering wiring system.
5. Not appropriate in the acidic, antacids and also moisture problems.

1.3.3. Lead Sheathed Circuit

The sort of circuit employs conductors that are protected with VIR and covered with an outer sheath of lead lightweight aluminum alloy having concerning 95% of lead. The metal sheath provided security to cords from mechanical damage, wetness as well as climatic deterioration.

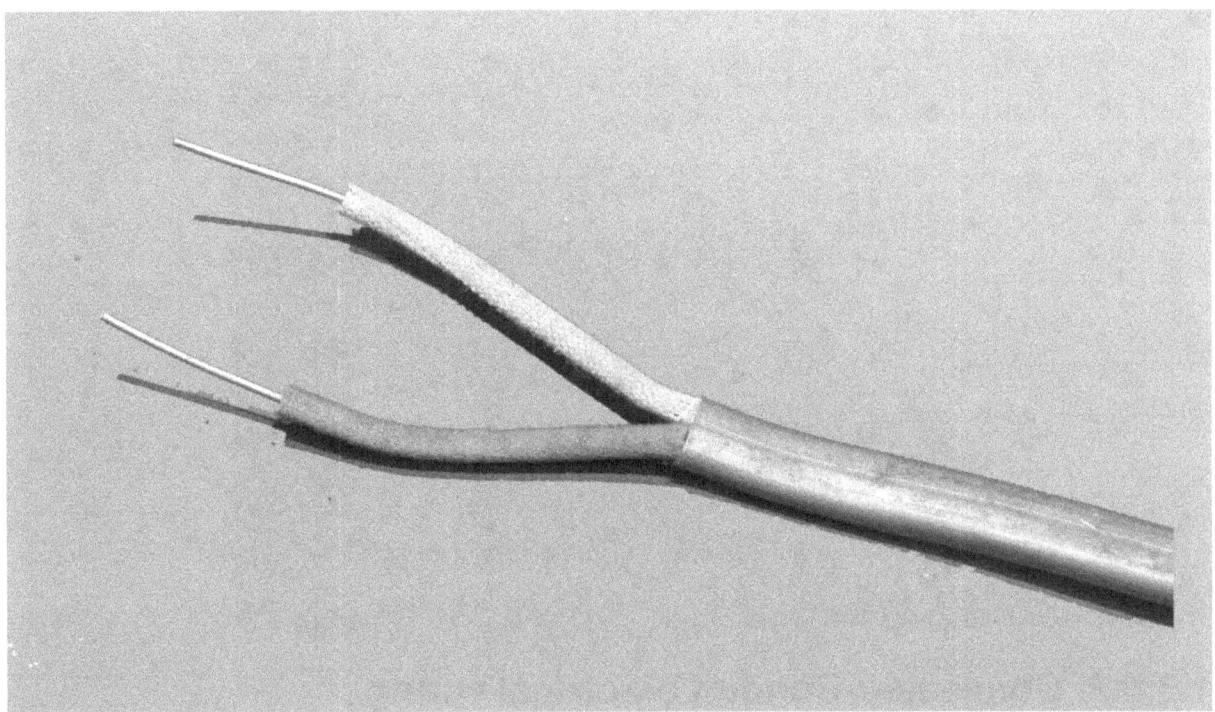

The entire lead covering is made electrically constant as well as is linked to earth at the point of access to shield versus electrolytic activity due to leaking existing as well as to give security in case the sheath becomes active. The cables are operated on wooden batten and also taken care of using web link clips just as in TRS electrical wiring.

1.3.4. Conduit Electrical wiring

There are two additional types of wiring according to pipe installment

a. Surface Conduit type
b. Concealed conduit type

1.3.4.1 Surface Conduit type

If channels set up on roofing or wall, It is referred to as surface conduit wiring in this wiring method, they make holes on the surface of wall on equivalent distances and also conduit is mounted after that with the help of rawal plugs.

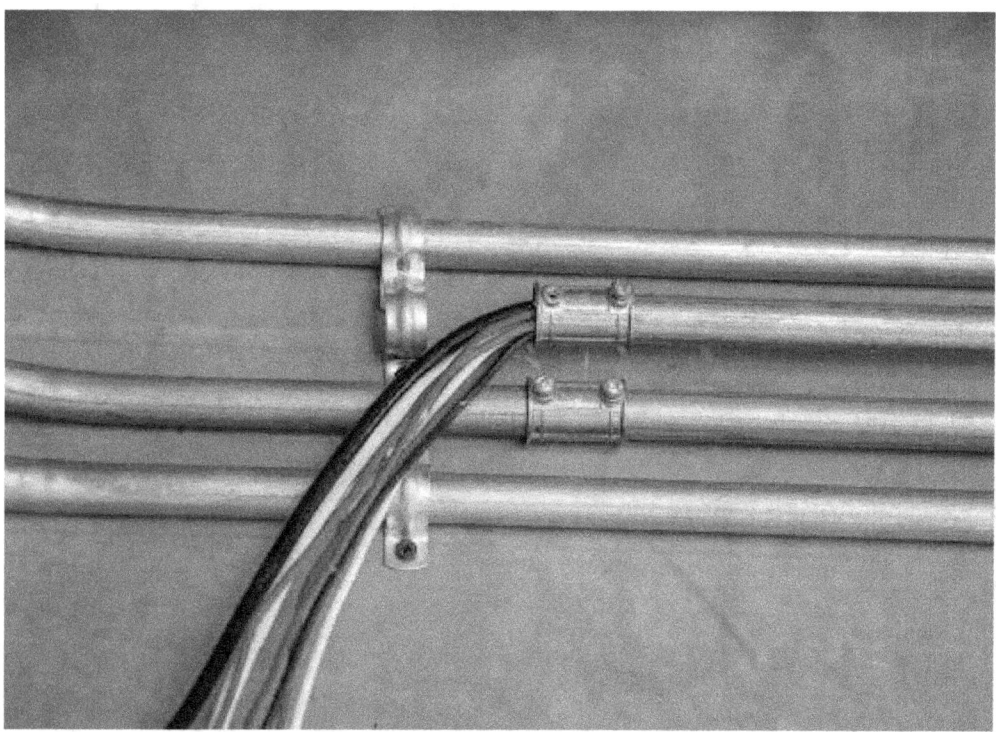

1.3.4.2. Concealed-Conduit electrical wiring.

If the channels are concealed inside the wall slots with the help of plastering, it is called hidden conduit circuit. In other words, the electric wiring system inside wall, roofing system or flooring with the help of plastic or metal piping is called hidden conduit wiring. obliviously, It is the most prominent, attractive, stronger and also typical electrical circuit system nowadays.

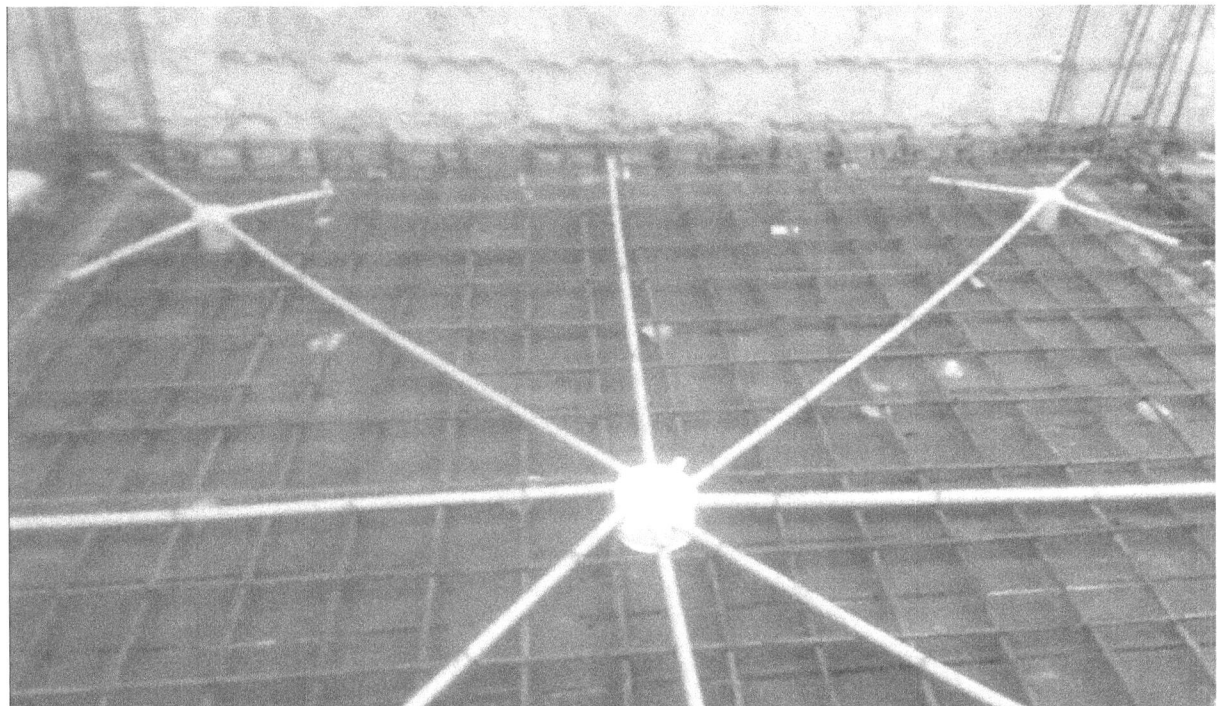

In conduit circuit, steel tubes called conduits are mounted externally of wall surfaces through pipe hooks (surface area conduit circuit) or buried in walls under plaster and also VIR or PVC cables are afterwards attracted by means of a GI wire of size if concerning 18SWG.

In Conduit wiring system, The channels need to be electrically constant as well as linked to planet at some suitable points in case of steel conduit. Conduit electrical wiring is a professional way of circuit a structure. Primarily PVC channels are made use of in residential electrical wiring.

The channel safeguards the cables from being damaged by rodents (when rats attacks the cords it will cause short circuit) that is why circuit breakers remain in place though but avoidance is better than treatment. When the structure is susceptible to fire mishap, lead channels are made use of in factories or. Trunking is more of like surface area conduit circuit. It's getting appeal too.

It is done by screwing a PVC trunking pipe to a wall surface then passing the cords via the pipe. The wires in conduit need to not be as well tight. Room aspect need to be put into consideration.

1.3.4.3. Sorts of Conduit

Complying with conduits are utilized in the channel circuit systems (both concealed as well as surface channel electrical wiring) which are shown in the above photo.

    a. Metal Channel
    b. Non-metallic conduit

a) Metal Conduit:

Metal conduits are made of steel which are pricey yet very solid too.

There are 2 sorts of metal channels.

    1. Class A Conduit: Low gauge conduit (Slim layer steel sheet conduit).
    2. Course B Channel: High scale conduit (Thick sheet of steel channel).

b) Non-metallic Channel:.

A solid PVC conduit is used as non-metallic channel now a days, which is easy and also versatile to bend.

C) Size of Conduit:.

The typical conduit pipes are readily available in various sizes genially, 13, 16.2, 18.75, 20, 25, 37, 50, and also 63 mm (diameter) or 1/2, 5/8, 3/4, 1, 1.25, 1.5, and also 2 inch in size.

1.3.4.4. Advantage of Conduit Circuit

1. Customization can be easily done according to the future requirements.
2. Fixing and also maintenance is very easy.
3. There is no risk of damages the wires insulation.
4. It is the most safe electrical wiring system (Hidden conduit wring).
5. Look is really lovely (in case of concealed channel circuit).
6. No threat of mechanical wears & tears and terminates in case of metallic pipes.
7. No danger of electrical shock (In case of proper earthing and grounding of metallic pipes).
8. It is dependable and also popular circuit system.
9. Long-lasting and also lasting electrical wiring system.
10. It is secure from rust (in case of PVC conduit) as well as danger of fire.
11. It can be used even in moisture, chemical impact and also smoky locations.

1.3.4.5. Drawbacks of Conduit Wiring

1. Installation is straightforward and also not very easy.
2. Threat of electric shock (In case of metal pipes without appropriate earthing & grounding system).
3. Extremely complicated to take care of added connection in the future.
4. It is costly wiring system (As a result of PVC as well as Metal pipes, Extra earthing for metal pipelines Tee and elbow joints etc.)
5. Very difficult to discover the issues in the electrical wiring.

If the channels are concealed inside the wall slots with the aid of plastering, it is called concealed conduit electrical wiring. In other words, the electrical wiring system inside wall surface, roof covering or floor with the help of plastic or metal piping is called concealed conduit wiring. In Conduit circuit system, The conduits need to be electrically constant and also connected to planet at some suitable factors in situation of steel conduit. Conduit electrical wiring is a professional means of wiring a building. Trunking is more of like surface conduit circuit.

2. Circuit Breaker selection and Calculation with an exercises

According to NEC (National Electric Code), IEC (International Electrotechnical Payment) and also IEEE (Institute of Electrical as well as Electronics Engineers), an appropriate rating of circuit breaker is must for all electric circuits i.e. property electrical wiring and industrial or commercial installment to stop the electrocution, dangerous fire as well as protection of the connected electric devices and also devices.

For optimum safety and security and trusted operation of the electric makers, it is advised to utilize the right and also suitable size of circuit breaker according to the circuit's current moving with it.

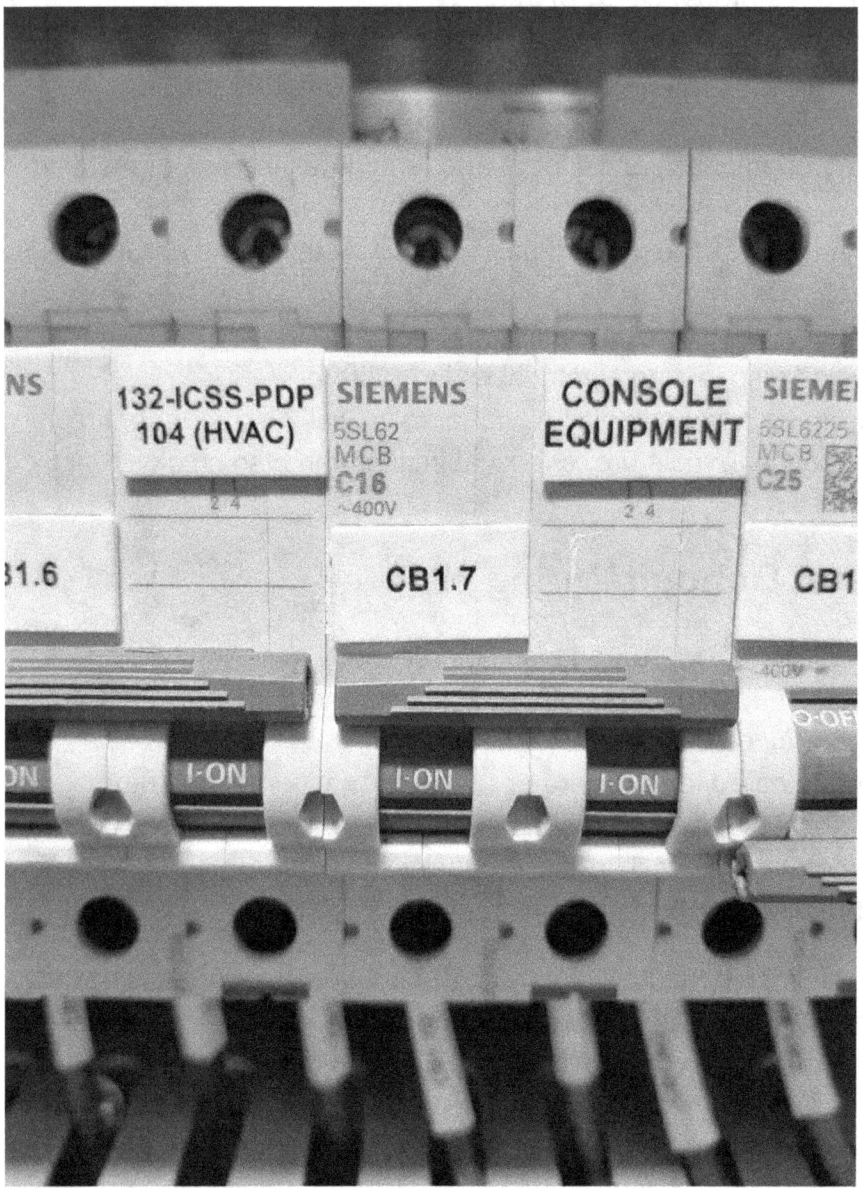

In case of various other (over or lower) size instead of right sized breaker, the circuit, cords and also cable also the linked gadget may heat up or in case of short circuit, it might start to smoke as well as melt. That's why a right size breaker is required for smooth procedure.

In this message, we will certainly be mosting likely to reveal that exactly how to choose a right size breaker for electric circuit installation and style with related voltage level, power level usage and also distinction in % to the circuit load as well as current rating of the CB.

2.1. What is a Breaker?

A Circuit breaker (CB) is a control and also protection tool which:

1. Break a circuit instantly under fault conditions (like over load, short circuit, etc).
2. Control (make or damage) a circuit by hand or by remote control under normal as well as fault conditions.

A circuit breaker is utilized for switching over device as well as protects the system

A breaker is a changing as well as protect gadget utilized for ON/OFF operation of the circuit as well as protect against the electrical shock. For accurate procedure and security, even complex styles are utilized with breaker like merges, passes on, switches, earthing & basing etc.

2.2. Working principle of Circuit Breaker

In regular problems when the circuit load rating is lower than the breaker rating, the circuit operation is regular and also it can be changed by hands-on procedure. In case of mistake or short circuit when the worth of current exceeds the circuit breaker current, It will immediately Trip i.e. Isolate the circuit from the main supply.

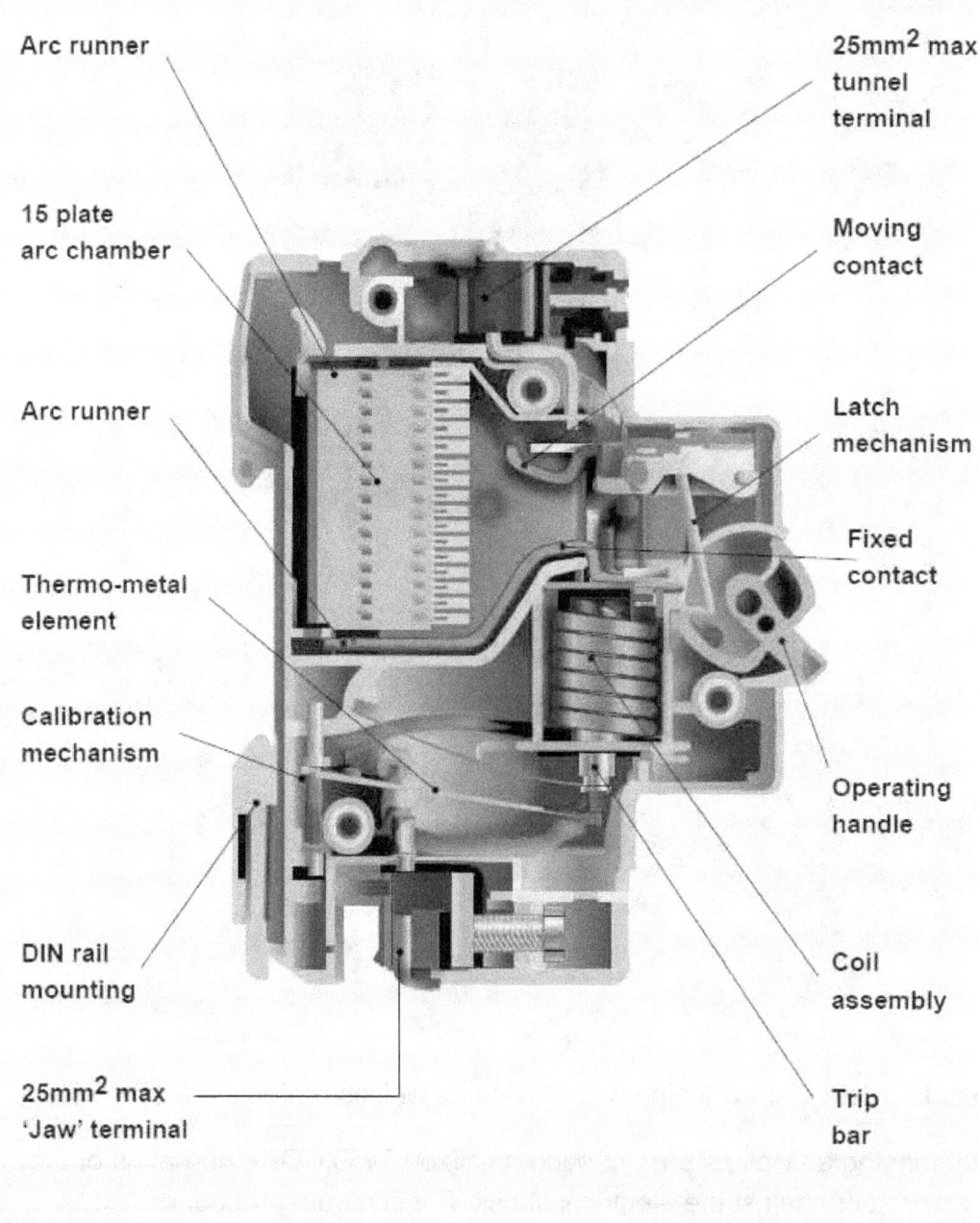

Arc runner

15 plate arc chamber

Arc runner

Thermo-metal element

Calibration mechanism

DIN rail mounting

$25mm^2$ max 'Jaw' terminal

$25mm^2$ max tunnel terminal

Moving contact

Latch mechanism

Fixed contact

Operating handle

Coil assembly

Trip bar

A 30 amp circuit breaker will trip at 30 amp no matter if, is it non or constant continual load. That's why we have to select 20-25% greater size of load for circuit breaker than the moving current in the cords as well as wires to the linked device.

If we use a 100A circuit breaker for 30A circuit, it wont protect the circuit from faulty currents as well as may damage the device and also shed as greater than 30A current will not trip the circuit breaker. Basically, we need to utilize the proper size of circuit breaker according to the device i.e. CB current need to not be reduced neither highest but 125% of circuit's load.

2.3. Single phase circuit breaker calculation.

To figure out the ideal rating of breaker for single ph supply, it depends on multiple aspects like type of load, cord material as well as environment temperature and so on. The basic general rule is that breaker rating must be 125% of the ampacity of cable-rating as well as the circuit which needs to be safeguarded by the CB. Allow see the following resolved examples:.

Example 1:.

Suppose, a 2 sq.mm wire is utilized for 18 amperes lighting circuit having 120V solitary stage supply. What is the very best rating of circuit breaker for that 18 A circuit?

Solution:.

Circuit Current: 18A.

Circuit Breaker Rating:?

CB size need to be 125% of the circuit existing.

= 125% x 18A.

= 1.25 x 18A.

Breaker Rating = 22.5A.

Example 2:.

What is the appropriate size of circuit breaker for 2000W, single stage 120V Supply?

Solution:.

- Load: 2000W.

- Voltage: 120V (Single Phase).

Circuit Current:.

According to the ohm's law,.

- $I = P/V$.
- $I = 2000W/120V$.
- $I = 16.66$ A.

Circuit Breaker Rating:.

Simply, Multiply 1.2 or 1.25 to the load current.

1.2 x 16.66 A.

Circuit Breaker Rating = 20 A.

Example 3:.

On 1840kW load, 230V single-phase circuit, find a suitable range of CB?

Solution:.

- Current = Power/ Voltage.
- $I = 1840W/230V$.
- $I = 8A$.

The minimal ranking of circuit breaker should be 8A.

The recommended rating of breaker need to be.

= 8A x 1.25.

= 10A.

2.4. Circuit Breaker Rating Estimation for 3 Phase Supply.

To locate the breaker rating for 3 phase supply voltage, we need to understand the exact load elements impacting the load current. To put it simply, exact same rule will not put on the different sorts of loads i.e. light, motor, capacitive or inductive loads as electric motor takes starting high current during the beginning procedure as well as power variable participation. For household usage, we may follow the same formula as over for single stage with taking the $\sqrt{3}$ (1.732) due to three phase power formula.

Great to recognize: For the exact same loads, the breaker rating in three phase is less than the breaker rating made use of in single phase ac circuits.

Allows locate the right size of circuit breaker for three stage circuits as comply with.

Instance 1: Which rating circuit breaker is needed for 6.5 kW, 3 phase 480V?

Answer:

Power in 3 Ph: P = V x I x $\sqrt{3}$.

Current: P/ V x $\sqrt{3}$.

- I = 6.5 kW/ (480V x 1.732) ... ($\sqrt{3}$ = 1.732).
- I = 6.5 kW/ 831.36.
- I = 7.82 A.

The recommended rating of breaker is.

1.25 x 7.82 A = 9.77 A.

The next closest criterion of circuit breaker is 10A.

Instance 2: Find the proper size of CB for 3-Phase 415V, 17kW?

Solution:.

- Current = Power/ (Voltage x $\sqrt{3}$).
- I = 17000W/ (415V x 1.732).
- I = 23.65 A.

2.5. Excellent to Know:.

- An over sized breaker used for protection can damage the water heater or various other linked home appliances also leads to the fire as a result of overheat.
- An undersized breaker or exact same ranking with load breaker can trip as well as reset the circuit running time and again. Utilize the right rating breaker.
- A single-ph breaker can not be used for 3 ph supply voltage degrees.
- A 3-Poles breaker can be made use of on 3-Phase system utilizing either 2 or 3 poles.
- If indicated by the markings or advised by individual handbook, - A 3-Poles circuit breaker can be used on 1-Phase system only and just.
- 30A Breaker and also 2.0 sq.mm can be used on 240V A/C Supply.
- Breaker can not be larger than ampacity of wire except for some loads like even more loads.

2.6. On top of that, A Breaker ranked for:.

- 120V can just be used for 120V.
- 240V can be made use of for 120V, 240V however not for 277V (Industrial applications).
- 120-277 can be utilized for 120V, 240V and also 277V.
- 120V can't be utilized on 240V circuit and the other way around.
- 15A-120V can't be used where in 20A, 120V circuit.

3. Exactly How to Wire Switches in Series and Parallel

3.1. Just How to Connect 2 Switches control in parallel on Single Load?

In previous fundamental residence electric circuit installation, we found out just how to wire single method switches in series. We will certainly learn just how to link as well as wire two Switches in parallel to regulate and run a single light.

Mainly, this is a favored technique to cord 1 way switches in parallel as series-parallel or parallel links are utilized alike electric circuit installment nowadays as a result of advantages over series link.

Before we enter details, we will certainly see the basic construction as well as running device of single way switch.

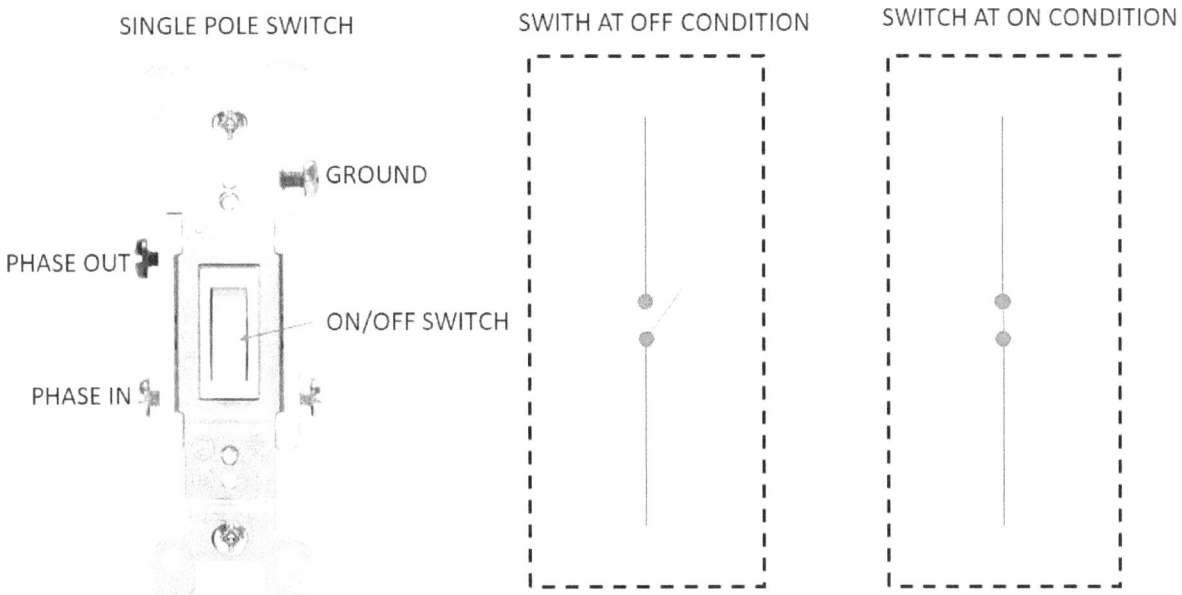

Construction & Working of one way SPST (Single Pole Solitary Through) Switch

Below is a straightforward detailed with schematic and circuit diagram which shows how to wire single method changes in parallel?

Needs:

- Solitary Method Switches (SPST = Single Pole Solitary Through) x 2 No
- Lamp (bulb) x 1 No
- Short pieces of loop wires x 5 No

3.1.1. Wiring Connection:.

Link the two single method switches, light bulb in parallel to the power supply as shown in fig below. One of the switches S1 or S2 need to be shut to complete the circuit.

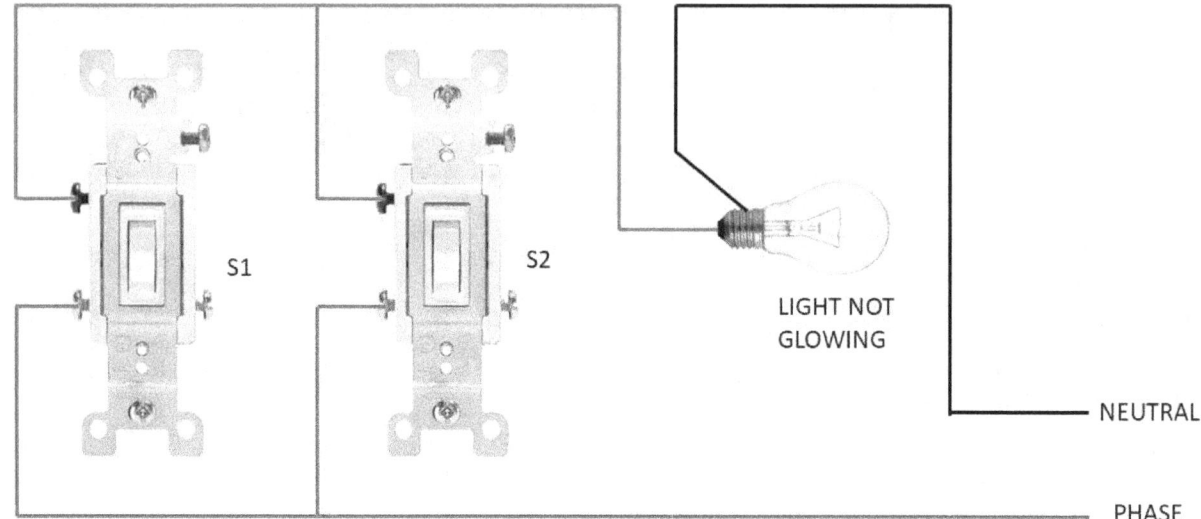

PARALLAL CIRUIT AND BOTH SWITCH OFF CONTION AND LIGHT NOT GLOWING

If there are more switches over connected in parallel with electrical appliance i.e. light point, one of them have to go to ON position to run the load. Nevertheless, light bulb will not go off if you turn off among the switch. In other words, all the switches have to be closed (OFF position) to separate the load from power supply.

3.1.2. Just How To Wires in Parallel?

The circuit will certainly complete if one of the switches out of 2 are at ON placement. In other words, If among the button are close or at ON placement, the light bulb will radiance after that. This is the same instance for various other loads as well to control by 2 (or more) single method switches linked in parallel.

When connected in parallel, Below are the various settings of single method switches as well as light point.

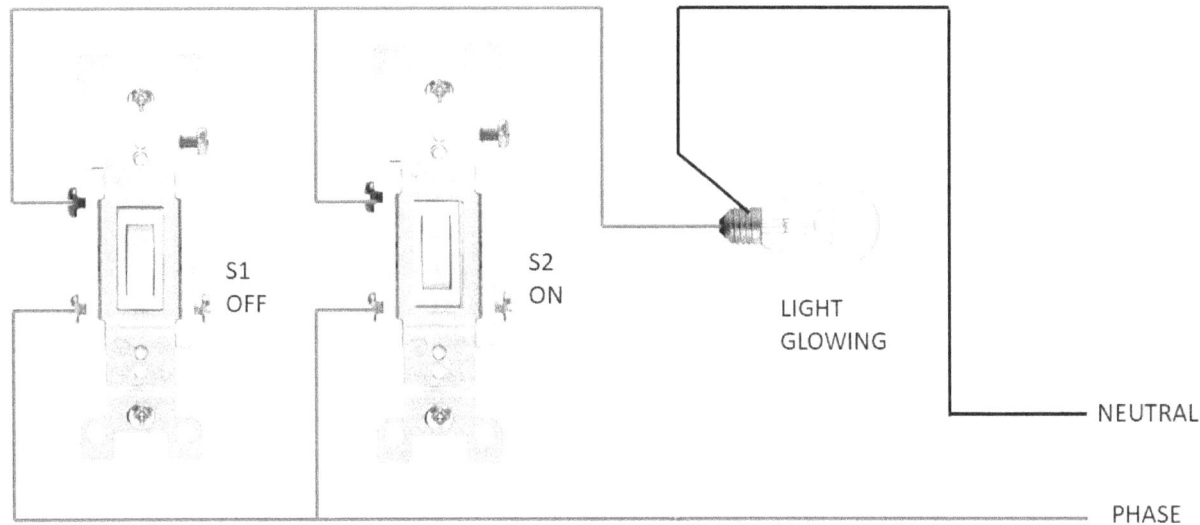

PARALLAL CIRUIT AND SWITCH 2-ON CONTION AND LIGHT GLOWING

Parallel switches in various Positions of Switches & Light Bulb

To get the switching setting in ON problem for light bulb, the above procedure is same as the Digital

Reasoning OR Gate truth table which is given below.

1st Switch	2nd Switch	Lamp status
OFF	OFF	OFF
ON	OFF	ON
OFF	ON	ON
ON	ON	ON

3.1.3. Great to know:

1. Fuses and Switches should be attached through line (Live) wire.
2. Switches over link in parallel is a like way to wire home devices. parallel or series-parallel electrical wiring method is much more trustworthy as opposed to series circuit.
3. Much more cables are needed in parallel-circuit link.
4. It is a comfy as well as trusted technique of circuit.

If there are much more switches over attached in parallel with electric home appliance i.e. light point, one of them should be at ON position to run the circuit. Light bulb won't go off if you switch over OFF one of the switch. In various other words, If one of the switch are close or at ON position, the light bulb will radiance after that. In easy words, there are 4 switching placements as well as if both the switches are at OFF position, the light bulb will not radiance. On the other hand, if one of the switch is at ON position, the existing will circulation in the circuit as the circuit behaves like finished circuit, hence light bulb will certainly glow.

3.2. Wire Switches in Series?

Just How to Attach 2 Switches in Series to Control a Solitary Load?

In fundamental house electrical circuit installment, we will certainly find out exactly how to wire as well as connect two switches in series to manage and also operate a single point.

Primarily, this is not a proffered method to cable solitary means switches in series as series-parallel or parallel connection is made use of alike electrical circuit installment. In some case, it might appears worthless connection, yet there are some possibilities where we have to regulate single load from 2 areas while both Switches should be activate to operate the circuit.

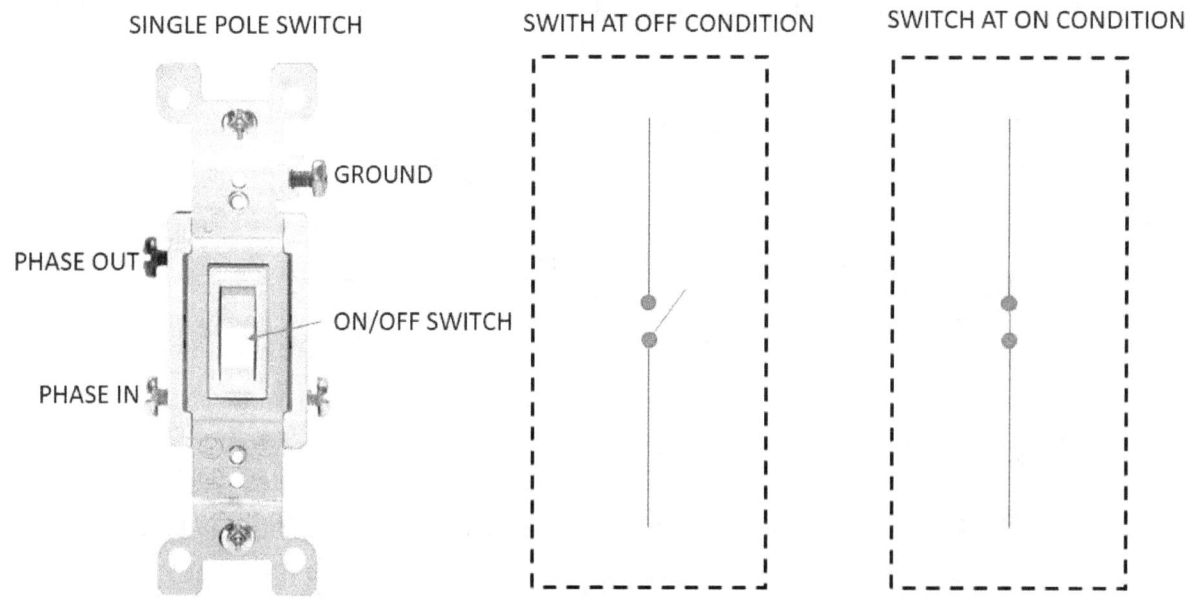

3.2.1. Construction & Working of one method SPST (Single Post Solitary With) Change

Below is a straightforward detailed with schematic as well as circuit representation which demonstrates how to wire single method switches over in series?

Demands:

- Solitary Way Switches Over (SPST = Single Post Solitary Via) x 2 No
- Lamp (Light Bulb) x 1 No
- Brief items of cords x 4 No

3.2.2. Working method

Attach both single way switches, light bulb in series to the power supply as shown in fig below. Both switches S1 as well as S2 should be shut to finish the circuit.

All of them have to be at ON setting to operate the load if there are more switches connected in series with electric home appliance i.e. light point. The circuit won't work after that if one of the single switch is open.

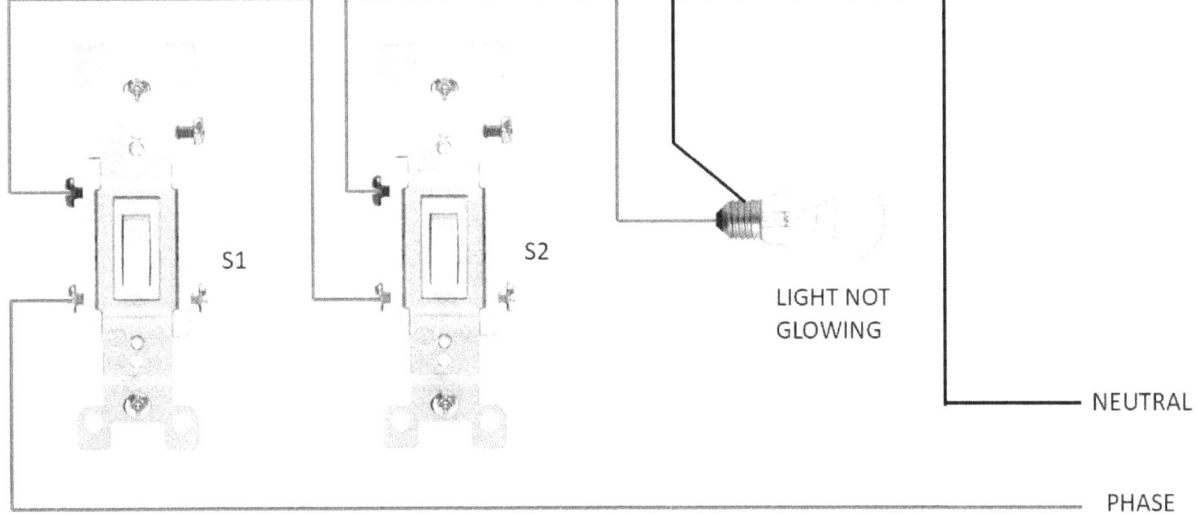

SERIES CIRUIT AND BOTH SWITCH OFF CONTION AND LIGHT NOT GLOWING

3.2.3. Just How to Wire Switches in Series?

The circuit will only complete if both of the switches go to ON setting. Simply put, If one of the button are open or at OFF placement, the light bulb will certainly not radiance. This coincides instance for various other load also which are attached in series to control by 2 single method switches.

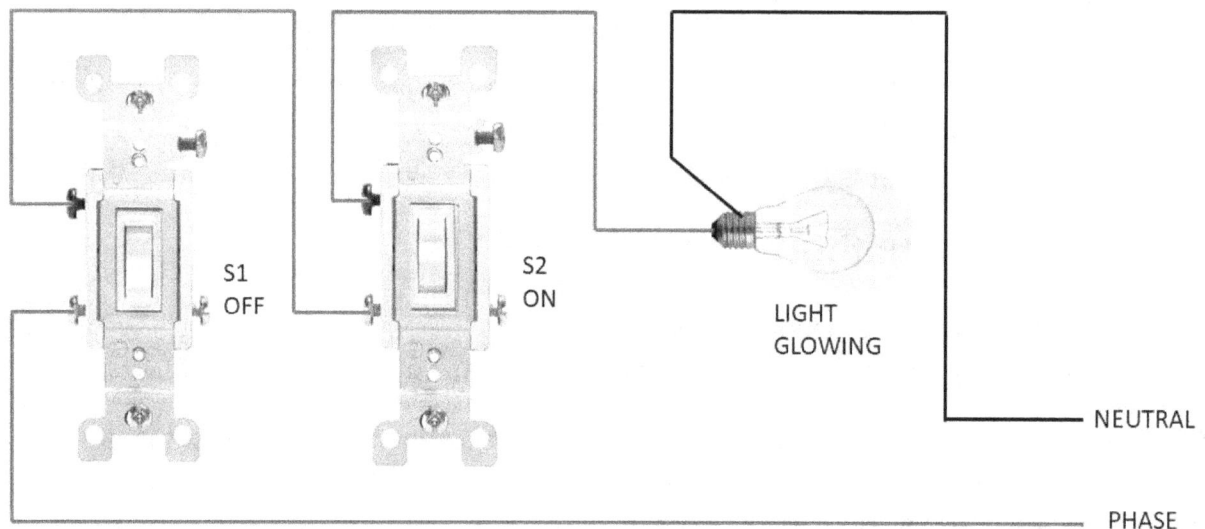

SERIES CIRUIT AND SWITCH-2 -ON CONTION AND LIGHT GLOWING

Logic for the series wiring in Different position of switch and light condition

To get the changing setting in ON condition for light bulb, the above procedure is like the Digital Logic AND Gate truth table which is given listed below.

Change 1Switch 2Lamp Position

1st Switch	2nd Switch	Lamp status
OFF	OFF	OFF
ON	OFF	OFF
OFF	ON	OFF
ON	ON	ON

In basic words, there are four switching settings as well as if both the switches are at ON setting, the light bulb will radiance. On the other hand, if one of the switch goes to OFF setting, the current will certainly not stream in the circuit as the circuit behaves like an open circuit, hence bulb will certainly not glow. Regardless of every one of various other linked Switches are at OFF or ON positions.

3.2.4. Excellent to understand:

  a. Switches as well as fuses need to be linked via line (Real-time) cord.
  b. Switches over connection in series are not a favor way to wire home appliances. parallel or series-parallel wiring approach is extra dependable.
  c. Much less wires and also cables are required in this kind of electrical wiring connection.
  d. It is not a comfy as well as reliable technique of electrical wiring.

The circuit will just complete if both of the switches are at ON position. In other words, If one of the button are open or at OFF placement, the light bulb will not glow. In straightforward words, there are four switching settings and if both the Switches are at ON setting, the light bulb will radiance. On the other hand, if one of the switch is at OFF setting, the current will certainly not move in the circuit as the circuit behaves like an open circuit, for this reason light bulb will certainly not radiance. No matter all of various other connected switches are at OFF or ON positions.

4. Regulate One Light from Two or Three Places by 2 Way Switch

Talk about 2-Way Switch

Two way changing link is made use of to regulate electric devices and also tools like Ceiling fans, lighting points etc from different areas by utilizing 2-way switches. One of the most common use 2-way changing link is stairs wiring where a light factor can be controlled from 2, 3 and even several areas. Whatever is the current position of 2 means switch (ON or OFF), the connected device like bulb can be activate/ OFF by pressing the switch.

4.1. Construction & Working of a 2-Way Switch over

2-Way switch is likewise referred to as Solitary Pole Double Through (SPDT). The fundamental construction as well as functioning concept of 2-way button is shown in (fig 1) below.

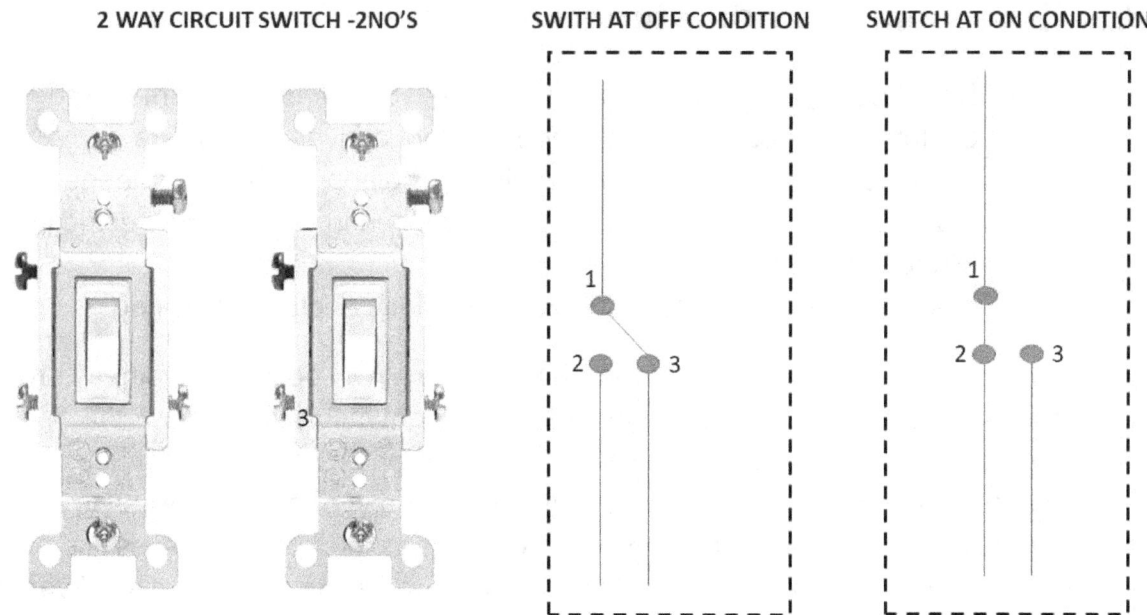

BREAKER OFF CONDITION: 1-2 NORMALLY OPEN
1-3 NORMALLY CLOSE

Construction & procedure of two means SPDT (Single Pole Dual Via) Switch

4.2. How to Wire a 2-Way Change

Below is a given schematic wiring layout that shows how to wire a 2-way switch and also regulate a light bulb from two various areas.

Note:

a. The same purpose can be attained by utilizing the following two method switching connection.
b. Connect the Earth Cord to the linked electric appliance as well as switches as per electrical regulation in your area.

4.3. Just How to Manage Light from Two Places by using 2-Way Switch over?

The adhering to two way switching connection can be made use of for the same function as reference over, i.e. to control a light point from two various places by using 2-way switches

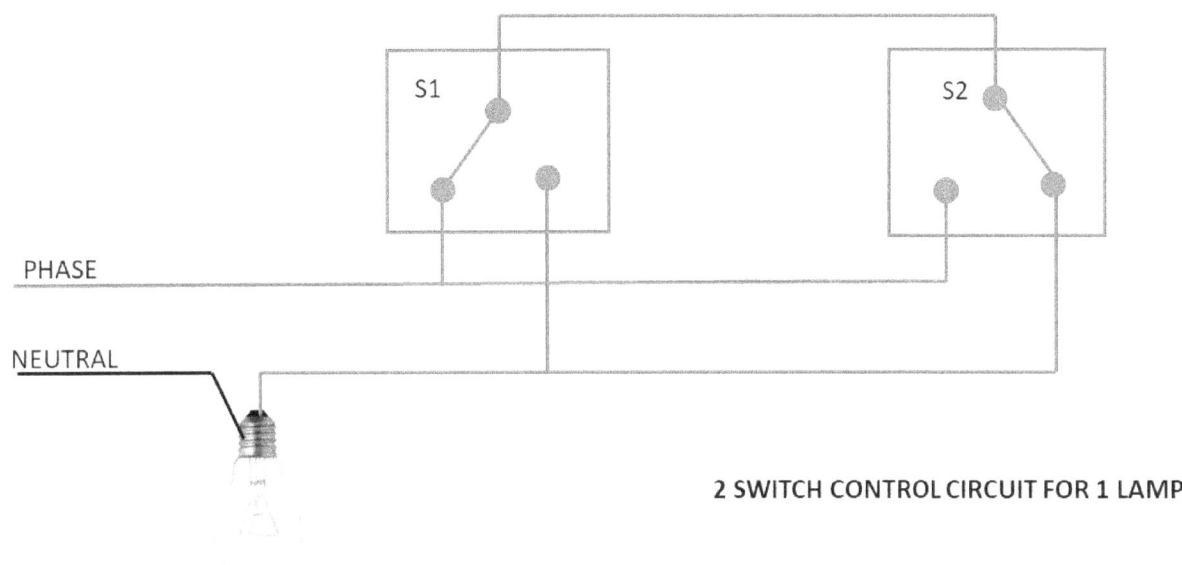

PHASE

NEUTRAL

2 SWITCH CONTROL CIRCUIT FOR 1 LAMP

4.4. Two Means switching to manage Light from Two Places in Stairs

As we went over above that the most usual use of 2-way Switches is to manage a light point from different areas like upper and downstairs i.e. reduced going into door and also upper door. This circuit is shown listed below:

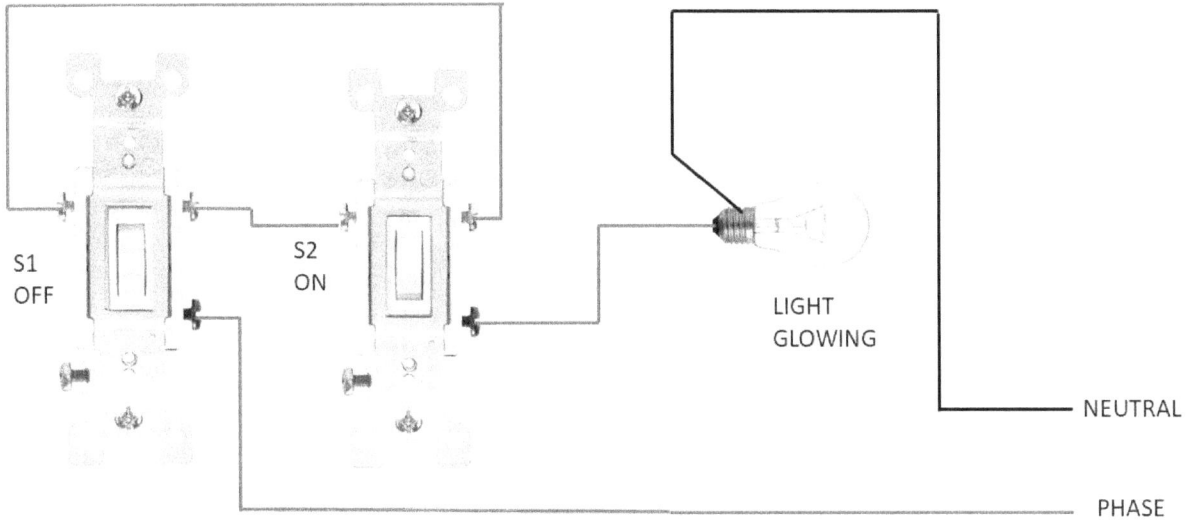

2-WAY CIRUIT, SWITCH-1 OFF AND SWITCH-2 -ON CONTION AND LIGHT GLOWING

4.5. Applications of 2-Way Switching

a. It is made use of to control electric tools as well as home appliances from two, three or perhaps a load more different places by including extra intermediate Switches.
b. It is additionally utilized in staircase electrical wiring link where a light factor can be control from 2 or even more different areas.
c. It is used in huge area rooms having two or even more entry and also exit doors and gates.
d. The main function of 2-way changing is to control an A/C or DC electric device, gadget or tools particularly light factors from 2 locations.

5. How to Regulate a Light Bulb by a Solitary Method or One-way Switch?

We will see how to wire a light switch to control a light point by one way circuit or single method button.

We will use the fundamental SPST (Single Post Single Through) switch in this tutorial to control a lamp/ light bulb from single place.

Before enter details, we have to show the fundamental of construction and operating system of solitary way switch which is displayed in fig below:

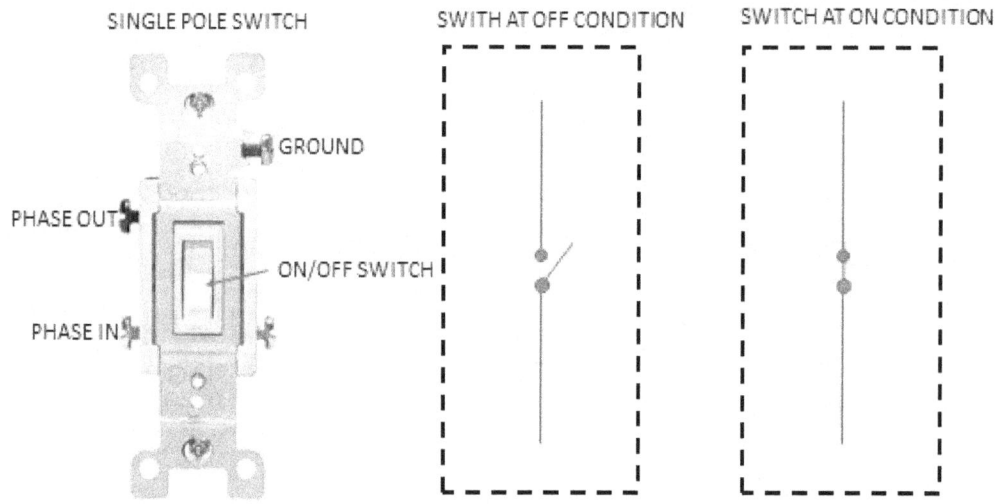

Below is an easy step by step tutorial with schematic as well as circuit diagram which shows how to wire a light button to regulate the bulb/lamp from single location with the help of one means or single means button?

Needs:

a. Solitary Means Change (SPST = Solitary Pole Single Through) x 1 No
b. Lamp (Light Bulb) x 1 No
c. Brief pieces of cable x 3 No

5.1. Procedure:

Fig given below shows the standard link of light button and also their placement i.e. when the button is OFF, the circuit acts like an open circuit and the light bulb won't glow. To switch over on the light bulb, switch S1 should be closed to finish the circuit as well as glow the light bulb.

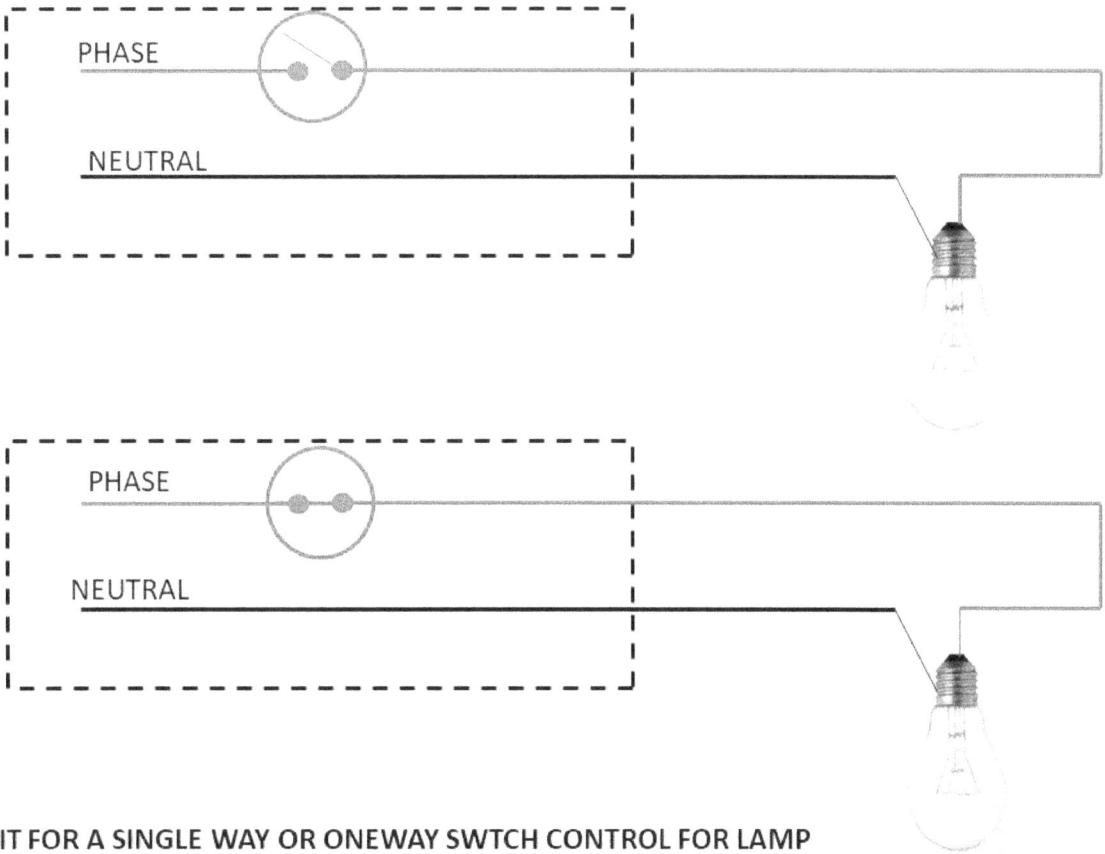

CIRCUIT FOR A SINGLE WAY OR ONEWAY SWTCH CONTROL FOR LAMP

5.2. Wiring for the Circuit

In fig listed below, schematic and also wiring representations of light switches are shown which demonstrates how to wire a light button?

This is simply like a series circuit i.e. all the parts are connected in series. Simply connect the Neutral cable directly to the light bulb and also then link the light bulb to the switch with middle cable. And also then attach the online cable to the button as revealed in the figures. Fig provided listed below programs the basic link of light button as well as their placement i.e. when the button is OFF, the circuit acts like an open circuit and the light bulb won't glow. To switch over on the bulb, button S1 have to be closed to finish the circuit and also glow the light bulb.

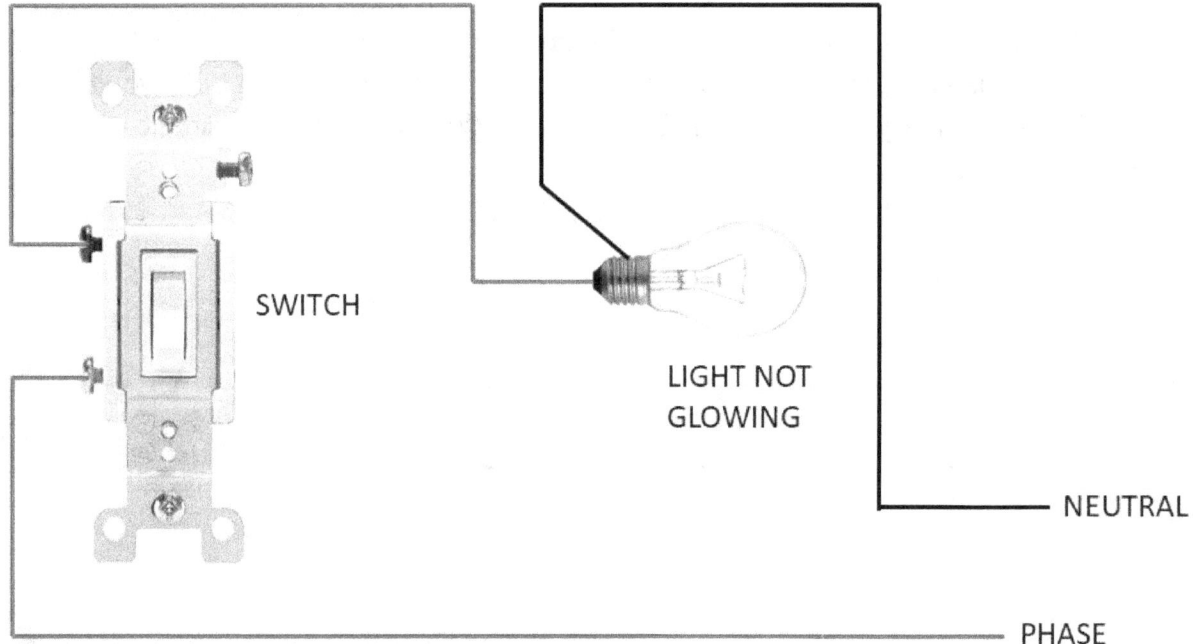

SWITCH

LIGHT NOT GLOWING

NEUTRAL

PHASE

CIRCUIT FOR A SINGLE WAY OR ONEWAY SWTCH CONTROL FOR LAMP

5.3. Just how to Wire a Light Change

Also note that house cable shades might vary according to different locations. Additionally, always use as well as attach the planet wire (direct naked cord to Switches, and electrical devices from planet web link in the circulation board to lower the danger of electric shock and also threat) which is disappointed in the numbers over.

5.4. Great to recognize:

a. Switches and also fuses need to be linked through line (Real-time) wire.
b. Changes connection in series is not a choose means to wire residence appliances. series-parallel wiring method is much more trustworthy.
c. Much less cables as well as cords are called for in this sort of electrical wiring connection.

6. How to connect Lights Points in Parallel?

The typical house circuits used in electrical wiring installation are (and need to be) in parallel. Mainly, Switches, Outlet receptacles and also light points etc are attached in parallel to keep the power supply to other electrical gadgets and also appliances through warm and also neutral cable in case if one of them gets fail.

6.1. Just How To Wire Lights in Parallel?

In the below fig, it is plainly shows that all the light bulbs are connected in parallel i.e. each light bulb attached with separate Line (additionally known as Real-time or Stage) and also Neutral cord.

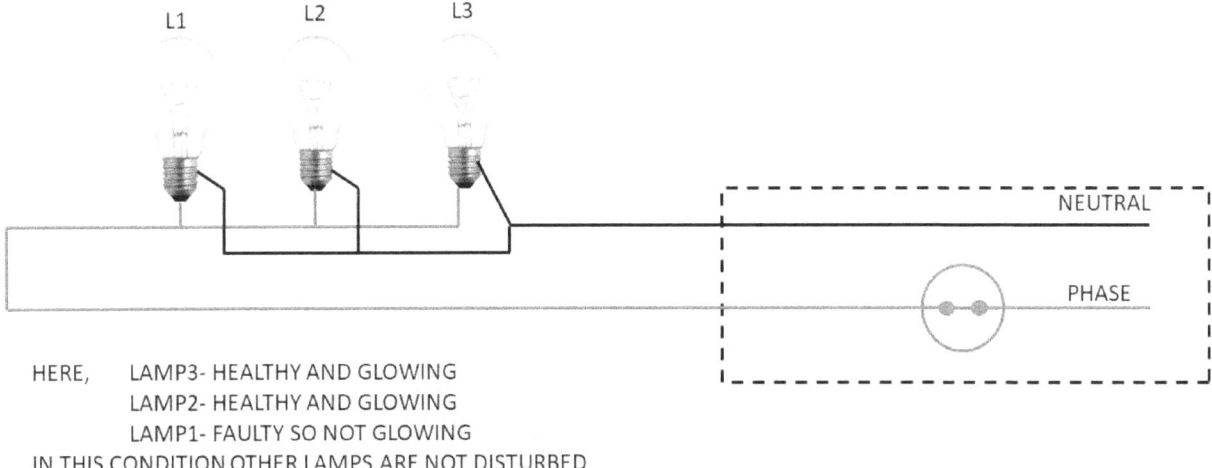

HERE, LAMP3- HEALTHY AND GLOWING
LAMP2- HEALTHY AND GLOWING
LAMP1- FAULTY SO NOT GLOWING
IN THIS CONDITION OTHER LAMPS ARE NOT DISTURBED

PARALLEL LIGHTS WIRING CIRCUITS

In parallel circuit, getting rid of or adding one light from the circuit has no effect on the others lights or connected gadgets and also appliances due to the fact that the voltage in parallel circuit is very same at each point yet the streaming current is different. Any type of number of lighting points or load can be included (according to the circuit or sub-circuit load calculation) in this sort of circuit by just extending the L and also N conductors to other lamps.

As each lamp or Bulb is attached between Line L and Neutral N independently, if among the light bulb gets faulty, the rest of the circuit will work smoothly as received fig below. Below, you can see there is a cut in the line wire linked to lamp 3, so the light bulb is switch OFF and the rest circuit is functioning effectively i.e. light bulbs are radiant.

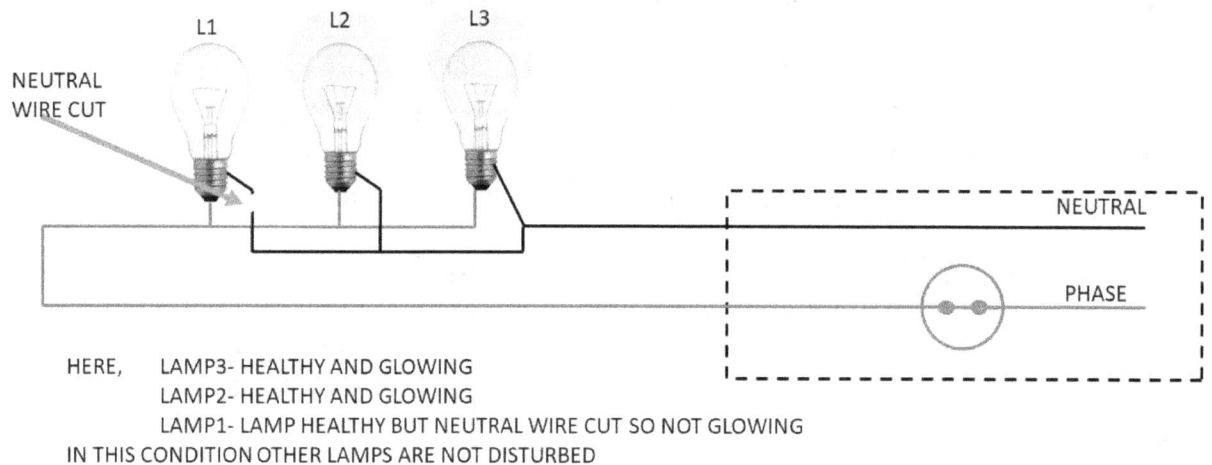

HERE,   LAMP3- HEALTHY AND GLOWING
         LAMP2- HEALTHY AND GLOWING
         LAMP1- LAMP HEALTHY BUT NEUTRAL WIRE CUT SO NOT GLOWING
IN THIS CONDITION OTHER LAMPS ARE NOT DISTURBED

PARALLEL LIGHTS WIRING CIRCUITS

6.2. Mistakes in Parallel illumination circuits

If we regulate each lamp by single way (SPST= Single Post Single Through) switch in parallel lights circuit, We will certainly be able to change ON/ OFF each light bulb from separate button or if we switch over OFF a bulb, the remainder lights factors won't impacted as is it occurs just in series lighting connection where all the linked load would certainly be detached if we close the switch.

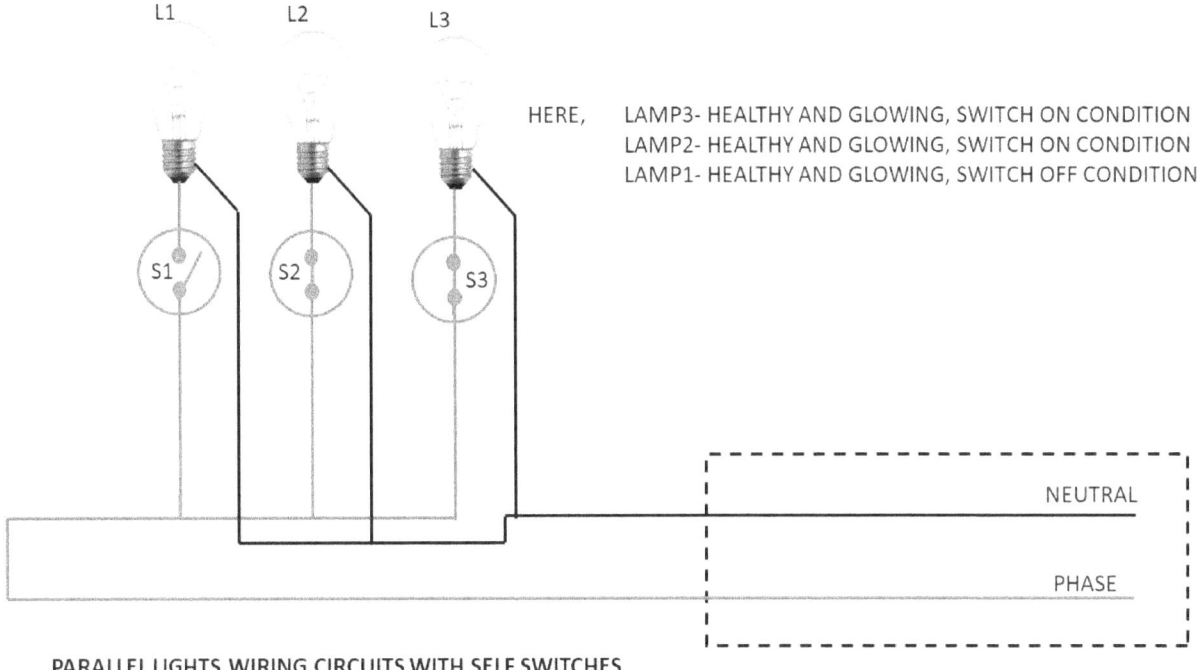

PARALLEL LIGHTS WIRING CIRCUITS WITH SELF SWITCHES

6.3. Light-Bulbs Connected in Parallel

Just How to Regulate Light Bulb from Single Way Change in Parallel Lights?

In below fig, We have managed 3 light bulbs from three different solitary means switches over attached in between line and neutral cords. The first two light bulbs are glowing as the switches go to ON position while the 3rd one bulb is turned off.

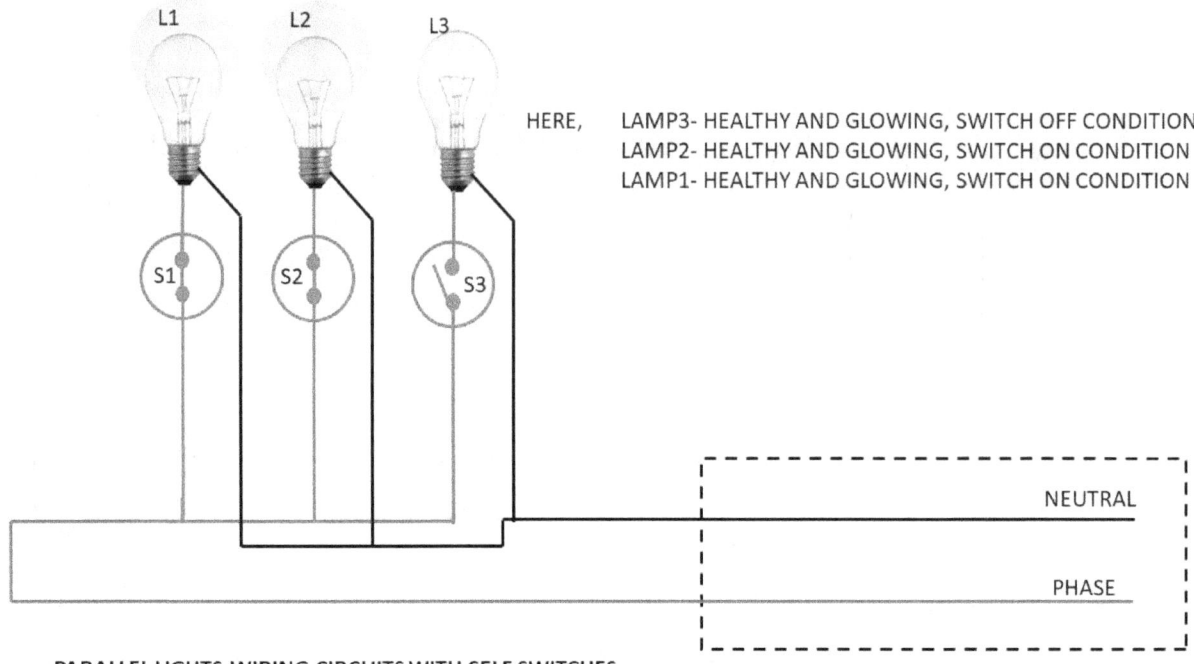

HERE,   LAMP3- HEALTHY AND GLOWING, SWITCH OFF CONDITION
LAMP2- HEALTHY AND GLOWING, SWITCH ON CONDITION
LAMP1- HEALTHY AND GLOWING, SWITCH ON CONDITION

PARALLEL LIGHTS WIRING CIRCUITS WITH SELF SWITCHES

Just how to control each lamp independently by single way changes in parallel lights circuits

6.4. Benefits of Parallel Lights Circuit:

1. Each connected electrical tool and home appliance are independent from others. By doing this, switching ON/ OFF a tool will not affect the other appliances and their procedure.
2. In case of break in the cord or removal of any type of lamp will certainly not break the all circuits and connected loads, simply put, other lights/lamps and also electrical appliances will certainly still work efficiently.
3. If even more lights are included the parallel lighting circuits, they will certainly not be reduced in brightness (as it takes place only in series lightning circuits). Due to the fact that voltage is exact same at each factor in parallel circuit. In other words, they get the exact same voltage as the resource voltage.
4. It is possible to include even more lighting fixture and lots points in parallel circuits according to future need regarding the circuit is not strained.
5. Adding additional devices and components won't boost the resistance yet will certainly reduce the general resistance of the circuit particularly

when high present ranking tools are utilized such as a/c unit and electric heating units.

6. Parallel wiring is more trustworthy, simple and also risk-free to make use of.

6.5. Drawbacks:

1. Much more dimension of cable as well as cable is used in parallel illumination electrical wiring circuit.
2. Much more existing needed when added light bulb added in the parallel circuit.
3. Battery runs out quicker for DC setup.
4. The parallel circuit design is extra complex as compare to series wiring.

6.6. Good to know:

1. Switches and also fuses should be connected via line (Real-time) cord.
2. Connecting electric gadgets and devices like fan, electrical outlet, light bulbs etc in parallel is a like means as opposed to series circuit.
3. Series-parallel or parallel wiring approach is more trustworthy rather than series wiring.

7. Lights in series Connection

We understand that series connection for house wiring like fan, Switches, light bulbs etc is not a recommended way instead of parallel or series-parallel circuit. In some situation, we require to cable and connect electric devices in series.

7.1. Just How To Wire Lights in Series?

In below fig, all the three light points are attached in series. Each light is connected to the following one i.e. the L (Line likewise known as online or stage) is attached to the initial light as well as other lights are connected through center wire and the last one cable as N (Neutral) attached to the supply voltage after that.

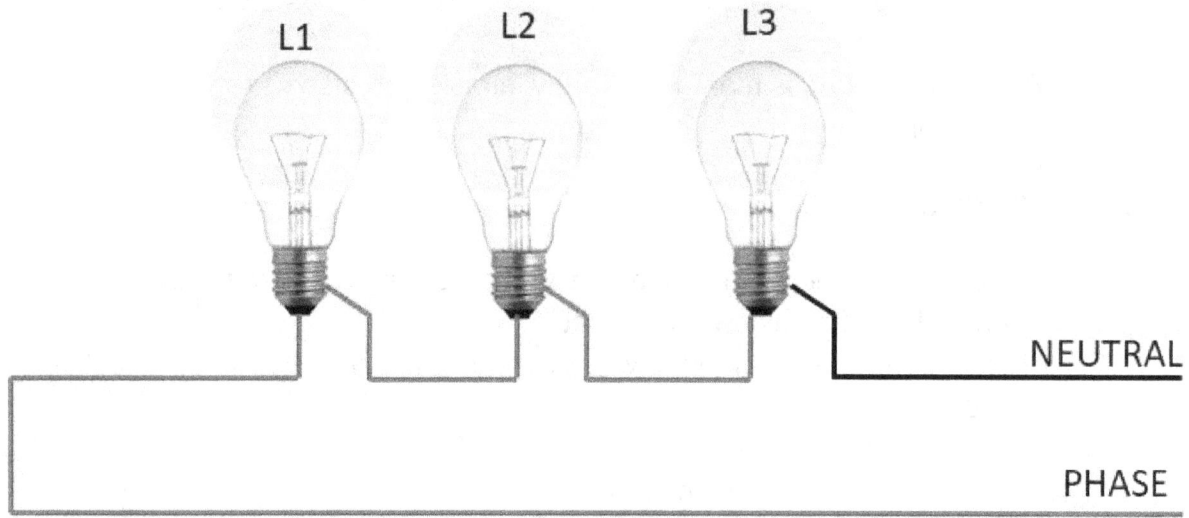

SERIES LIGHTS WIRING CIRCUITS

according to the series circuit analogy, the flowing current is same in all these incandescent bulbs/ lights but the voltage are various as opposed to the parallel circuit where voltage are exact same at each factor where existing are various.

Among the major disadvantage of series lights circuit, eliminating or adding one lamp from the circuit will certainly influence the over all circuit i.e. others lamps will lower in light and other linked devices as well as appliances will certainly not obtain the sufficient or needed run voltage since the voltage in series circuit is various at each factor but the moving current is exact same.

Any kind of number of lighting points or loads can be included (according to the circuit or sub-circuit load estimation) in this sort of circuit by just extending the L as well as N conductors to various other lamps yet they will certainly not radiant according to the rated outcome efficiency. simply put, including extra light bulbs in series circuit will lower the rest of the light points.

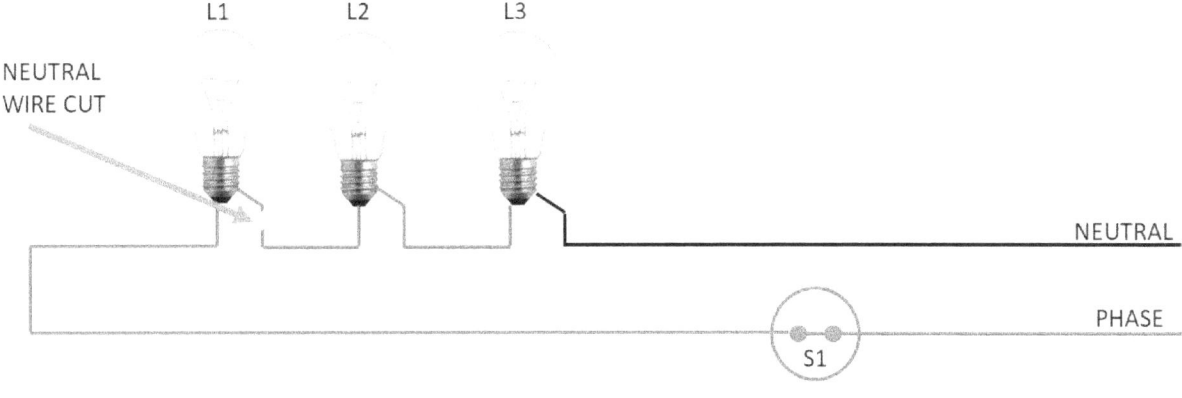

SERIES LIGHTS WIRING CIRCUITS

HERE, LAMP1- NOT GLOWING DUE TO L1 NEUTRAL WIRE CUT
LAMP2- NOT GLOWING DUE TO L1 NEUTRAL WIRE CUT
LAMP3- NOT GLOWING DUE TO L1 NEUTRAL WIRE CUT

An additional major problem of series lights circuit is that as all bulbs or lights are attached in between Line L and Neutral N appropriately, if one of the light bulb gets faulty, the rest of the circuit will certainly not function as the circuit will certainly be open as displayed in fig listed, you can see there is a cut in the line wire connected to light 3, so the light bulb is switch OFF and the rest circuit is functioning appropriately i.e. light bulbs are radiant.

7.2. Lights linked in Series Benefits:

a. Less size of cord cable is required in series wiring.
b. We use to protect the circuit to attach fuse & circuit breakers in series with other home appliances.
c. Series circuit don't obtain overhead quickly because of high resistance when more load added in the circuit.
d. The life expectancy of battery in series circuit is extra as compared to parallel.
e. It is most simple method of electrical circuit and also mistake can be conveniently identify and also repair as compared to parallel or series-parallel circuit.

7.3. Disadvantages of Series Illumination Circuit.

a. The break in the wire, failing or elimination of any kind of single lamp will certainly break the circuit and trigger every one of the others to quit working as there is just one single course of current to flow in the circuit.

b. If even more lamps are included series illumination circuit, they will certainly all be lowered in illumination. Due to the fact that voltage is cooperated series circuit. If we add extra load in series circuit, the more than voltage decline is rises which is not a good sign for electric appliances security.
c. Series Wiring is "ALL or NONE" type circuit imply all the appliances will operate at when or all of them will separate if mistake occurs at any among the connected tool in series circuit.
d. High supply voltage are required if we require to add even more load (light bulbs, electrical heating units, a/c etc) in the series circuit. If five, 220V bulbs are connected in Seriesly, Then circuit Voltage shall be: 5 x 220V = 1100V.
e. The overall series circuit resistance boosts (and also present reductions) when even more load added in the circuit.
f. According to future need, just those electric devices need to be added in the present series circuit if they has the same existing ranking as existing are exact same at each point in series circuit. We recognize that electrical appliances and also devices i.e. light bulbs, fan, heating unit, air conditioner etc have different present score, for that reason, they can not be linked in series circuit for efficient as well as smooth operation.

7.4. Excellent to know:

a. Switches and also integrates need to be attached via line (Real-time) wire.
b. Linking electrical devices and appliances like fans, outlet, light bulbs etc in parallel is a like way instead of series circuit.
c. Parallel or series-parallel wiring method is more trusted as opposed to series wiring.

8. Stairs Circuit Circuit Representation-- Exactly How to Regulate a Lamp from 2 Places by 2-Way Switches?

In fundamental electric circuit setup, we will certainly review step by step technique of staircase electrical wiring installment by using 2-way switches (SPDT = Solitary Post Dual Via Change). The same wiring circuit diagram can be utilized for 2-way lights or regulating electric devices from 2 different places by using two-way Switches. The primary objective of two method switching circuit is that the appliances can be ON/ OFF separately from any kind of button, despite whatever is the present placement of the button.

8.1. Staircase Electrical Wiring Circuit Layout Link

Below is traditional stairs wiring circuit representation. Right here we can manage a bulb from 2 different locations by utilizing 2 2-way switches.

Working & Operation of Stairs Circuit-- 2-Way Light Switching

Take into consideration the above 2-way button wiring diagram which has been used to control a bulb in staircase. The schematic shows that circuit is completed and light bulb gets on. Suppose you want to OFF the light bulb from the top button at top of staircase (upper section of stairs) simply Switch OFF the button then circuit will certainly damage and the bulb will certainly be OFF. To activate the bulb once more, just activate the same switch at top portion of staircase. Simply put you can On And Off light bulb from upper button at the top of staircase. Obviously; you can execute the very same operation from the bottom switches mounted in stairs

Now, allows see just how we can do that from the various other button installed at the end of stair.

For this function, take into consideration the figure given over. In this situation, you can see that circuit is total and also light bulb gets on. Intend you want to OFF the light bulb from the reduced button at bottom of staircase. Merely OFF the switch, however circuit will damage and also the light bulb will be OFF. You can switch ON the light bulb once more to activate the same switch mounted near the bottom or downstairs as shown in the fig.

8.2. Two-way Switching Control making use of Three Wires

It is the 2-way changing link technique which can be used for staircase circuit as well as it is efficient as compared to the old-school technique where 2 wires are made use of as opposed to 3 wires.

The current placement of 2 method switching link making use of three cables circuit is ON and the bulb is beautiful. The circuit operation is same as mentioned in the above numbers however the connection approach is different as the initial terminals of both switches are linked to the online (Phase) cord. The 2nd terminals of both switches are linked to the light bulb to offer online line supply while the Neutral is straight attached to the bulb as common electrical wiring technique.

This fundamental circuit is little complex as by looking in it, it makes a short circuit (to the exact same wire which is not hurt in this instance) when both Switches are ON or OFF which is making a loophole to disconnect the live supply to the bulb, Therefore, bulb will certainly not glow in that situation.

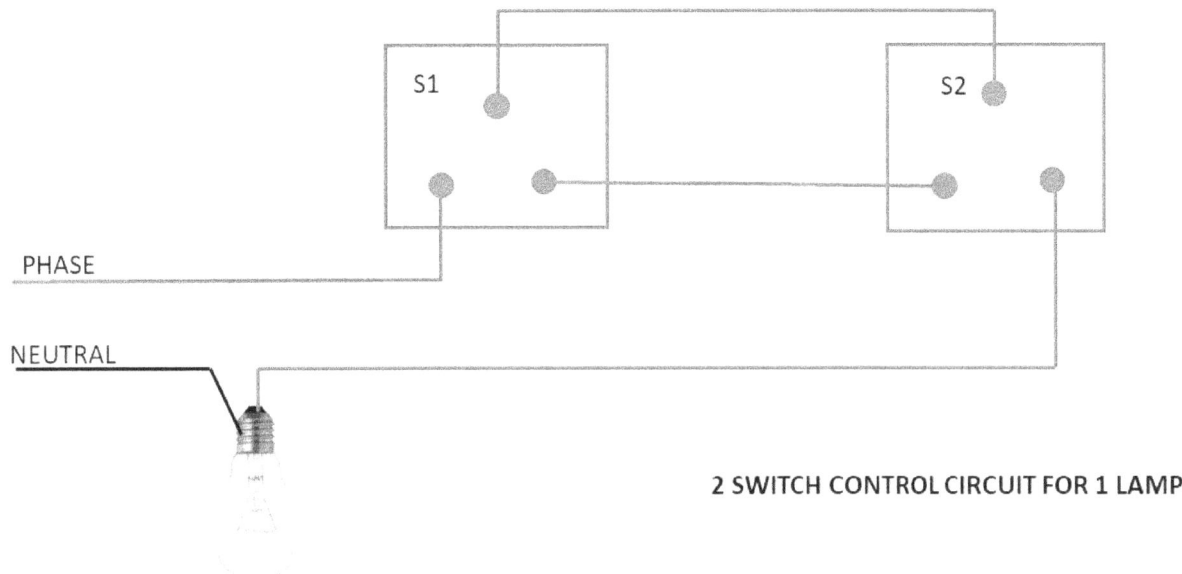

2 SWITCH CONTROL CIRCUIT FOR 1 LAMP

To get the changing placement know problem for light bulb, the above operation is same as the Exclusive-OR (EX-OR) logic gateway truth table which is provided listed below.

1st Switch	2nd Switch	Lamp status
OFF	OFF	OFF
ON	OFF	ON
OFF	ON	ON
ON	ON	OFF

CONDITION-1

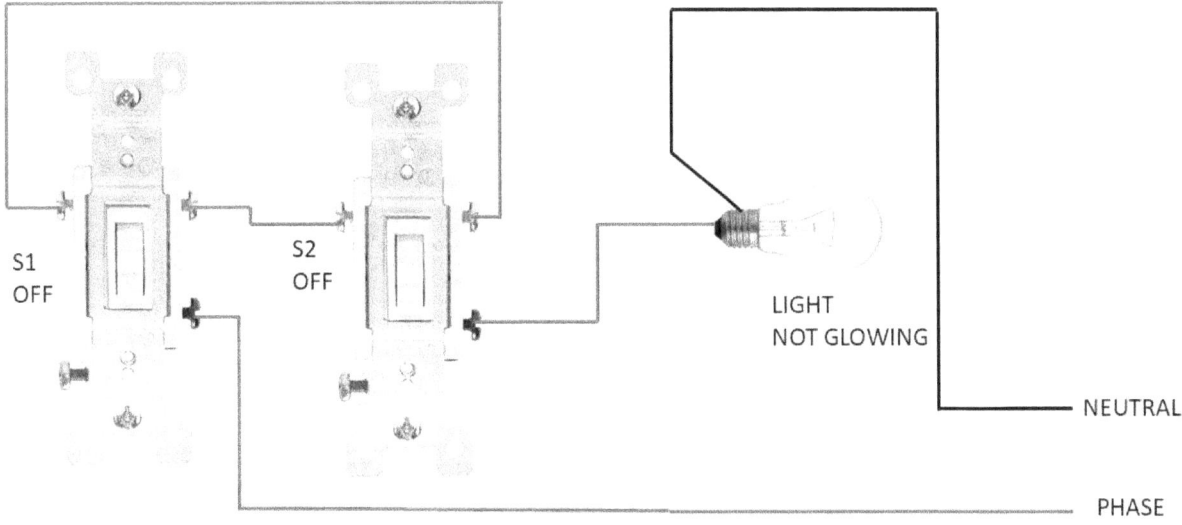

2-WAY CIRUIT, SWITCH-1 OFF AND SWITCH-2 -OFF CONTION AND LIGHT NOT GLOWING

CONDITION-2

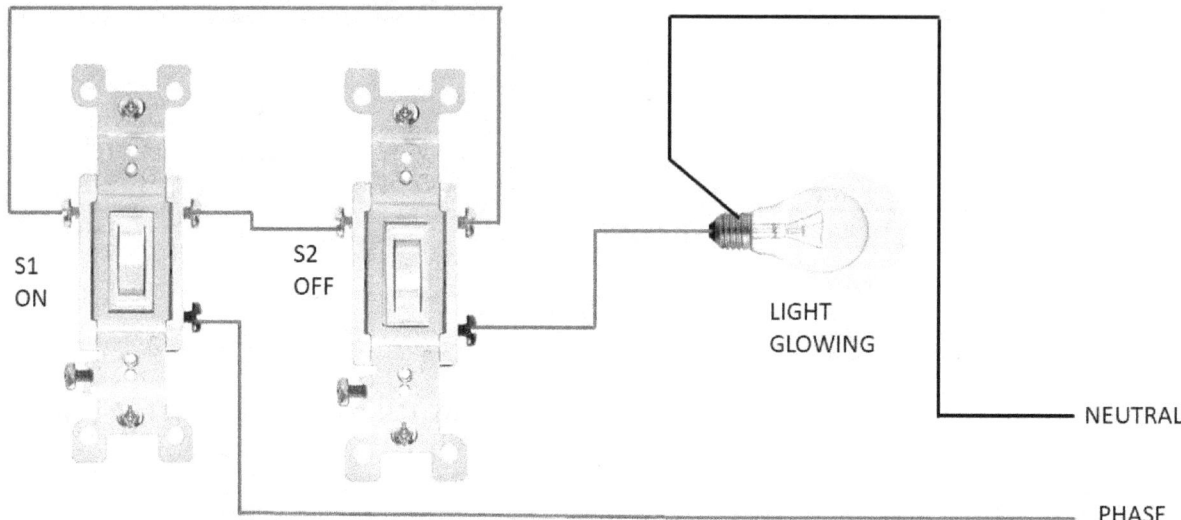

S1
ON

S2
OFF

LIGHT
GLOWING

NEUTRAL

PHASE

2-WAY CIRUIT, SWITCH-1 ON AND SWITCH-2 -OFF CONTION AND LIGHT GLOWING

CONDITION-3

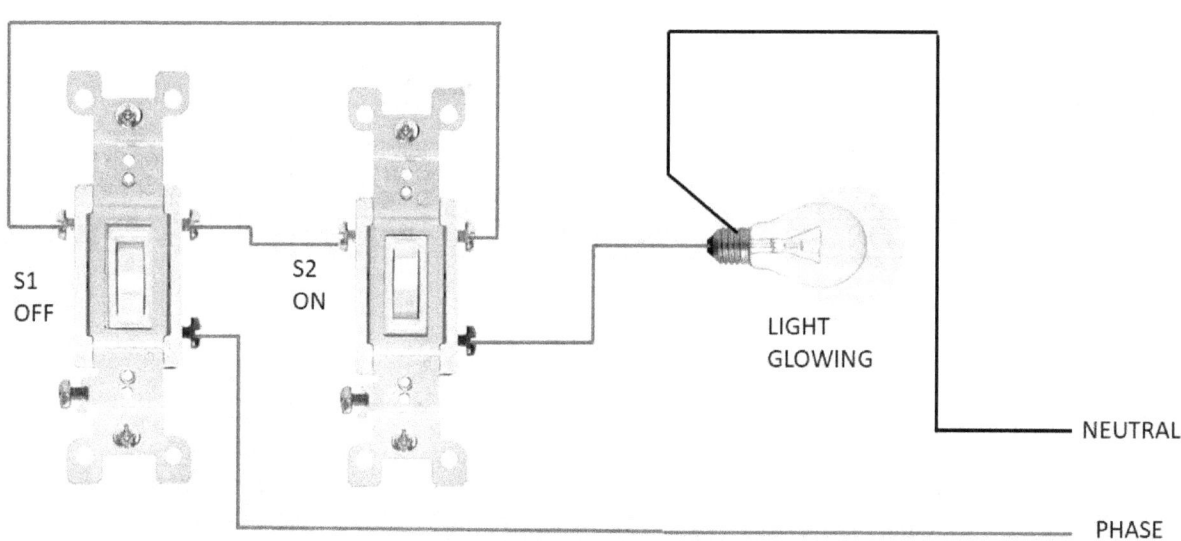

S1
OFF

S2
ON

LIGHT
GLOWING

NEUTRAL

PHASE

2-WAY CIRUIT, SWITCH-1 OFF AND SWITCH-2 -ON CONTION AND LIGHT GLOWING

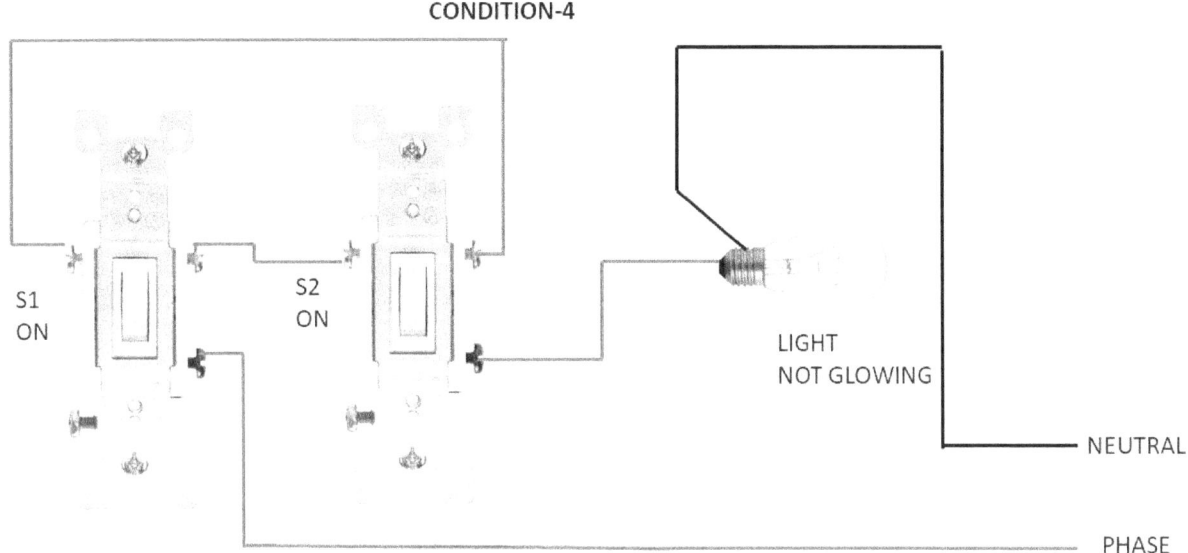

CONDITION-4

S1
ON

S2
ON

LIGHT
NOT GLOWING

NEUTRAL

PHASE

2-WAY CIRUIT, SWITCH-1 ON AND SWITCH-2 -ON CONTION AND LIGHT NOT GLOWING

9. Just how to Manage Each Lamp by Separately Switch in Parallel Lighting Circuit?

'How to wire and control each lamp independently by using separate solitary method switches in parallel lights link.'

Below is a simple step by step with schematic as well as circuit layout which shows how to wire three different light bulbs in parallel to control from 3 different as well as independent areas and Switches?

Needs:

1. Single Way Switches Over (SPST = Solitary Post Solitary Through) x 3 No
2. Lamp (Light Bulb) x 3 No
3. Short items of cords x 11 No

9.1. Procedure:

Link all the circuit link as displayed in fig listed below.

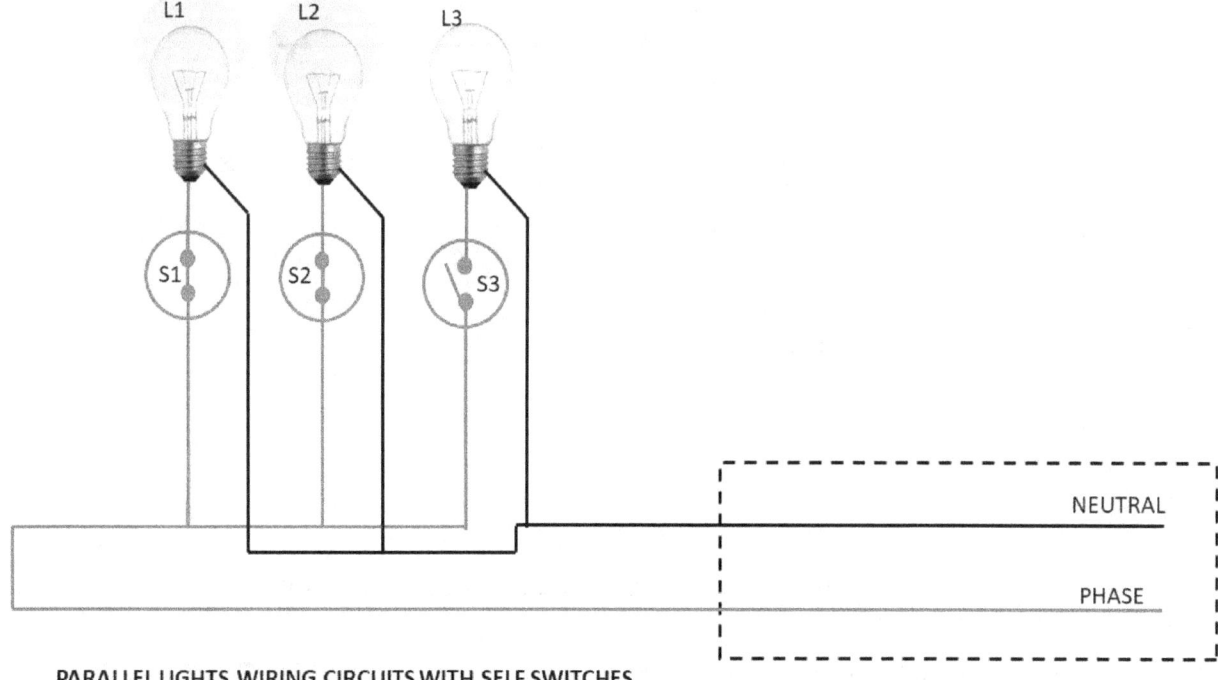

PARALLEL LIGHTS WIRING CIRCUITS WITH SELF SWITCHES

9.2. Just how to control each lamp individually by solitary way switches in parallel illumination circuits

a. The first and also 2nd Lights are glowing, due to the fact that both the separate switches S1 and S2 which are link to the light bulbs through Line are at ON setting therefore the circuit is completed.

b. The third Light is OFF, due to the fact that button S1 which is attached to light bulb via Line is OFF, so the circuit acts like an open circuit meaning there is no other way to flow the present in the circuit. Thus, light bulb is not radiant

Currently think about the adhering to schematic electrical wiring representation. It is the same circuit as revealed above but the Switches as well as light bulb delays simply reversed i.e. S1 and S2 are at OFF positions, so Light 1 and Light 2 are OFF while S3 is ON and also Lamp 3 is radiant.

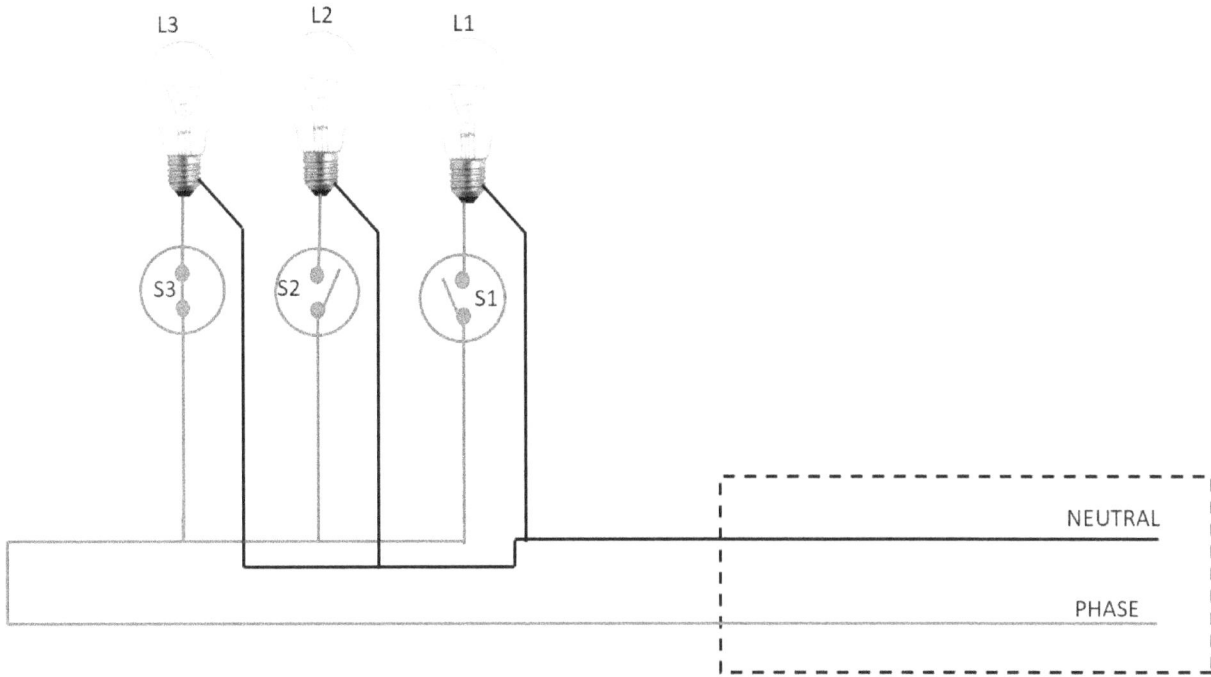

PARALLEL LIGHTS WIRING CIRCUITS WITH SELF SWITCHES

Light Bulbs Connected in Parallel

9.3. Great to know:

a. Switches and also integrates need to be attached via line (Online) wire.
b. Linking electrical tools and also appliances like fans, outlet, light bulbs etc in parallel is a like method as opposed to series wiring.
c. Parallel or series-parallel circuit approach is a load can be more worth instead of series wiring.

10. Which Bulb Glows More Vibrant When Connected in Series as well as Parallel & Why?

2 Light Bulbs of 80W & 100W are linked in Series & Parallel-- Which One will Glow Brighter

One of the most complicated concern we obtained that if 2 bulbs are linked in series and after that in parallel, which one will radiance brighter and what are the exact reasons? Well, there are great deals of info around the internet, but we will certainly go in very detailed information to calculate the exact worths to get rid of the confusion.

Of all, keep in mind that the bulb having a high resistance and dissipate even more power in the circuit (no matter series or parallel) will radiance more vibrant. In other words, the brightness of the bulb depends upon voltage, existing (V x I = Power) in addition to resistance.

The light brightness is directly proportional to the bulb power level. That's why the even more power level a light bulb is using will certainly glow brighter.

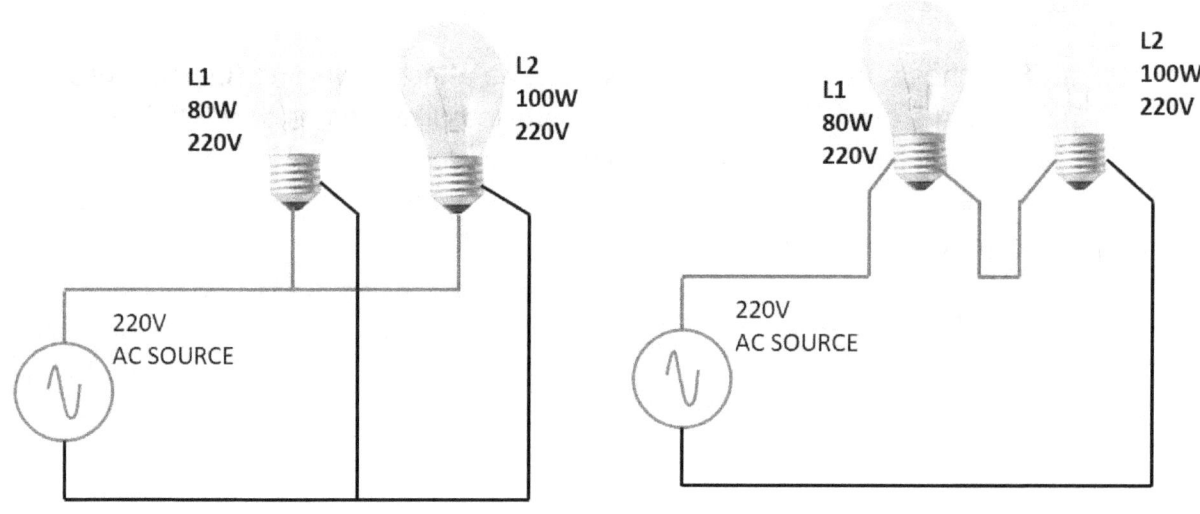

WHICH CAN BE MOST BRIGHTER 80W OR 100W AT
SERIES AND PARALLAL CONNECTION

10.1. When Light bulbs are connected in Series

Rankings of light bulbs Wattage are different and linked in a series circuit:

Expect we have 2 light bulbs each of 80W (Bulb 1) as well as 100W (Bulb 2), ranked voltages of both bulbs are 220V as well as connected in series with a supply voltage of 220V A/C. In that instance, the bulb with high resistance and also even more power dissipation will radiance brighter than the other one. In short, In series, both light bulbs have the same current flowing through them.

Power

$P = V \times I$ or $P = I2 R$ or $P = V2/R$

Currently, the resistance of Bulb 1 (80W);.

We understand that existing is same as well as voltage are additive in a series circuit but the rated voltage of bulbs are 220V. i.e.

Voltage in series circuit: $VT = V1 + V2 + V3 ...+ Vn$.

Existing in series circuit: $IT = I1 = I2 = I3 ... In$.

For that reason,.

$R = V2/ P80$.

$R80W = 2202/ 80W$.

$R80W = 605\Omega$.

And, the resistance of Light bulb 2 (100W);.

$R = V2/ P100$.

$R100W = 2202/ 100W$.

$R100W = 484\Omega$.

Currently, Existing;.

$I = V/R$.

$= V/ (R80W + R100W)$.

$= 220V/ (605\Omega + 484\Omega)$.

I = 0.202 A.

Currently,.

Power dissipated by Bulb 1 (80W).

P = I2R.

P80W = (0.202 A) 2 x 605Ω.

P80W = 24.68 W.

Power dissipated by Light bulb 2 (100W).

P = I2R100.

P100W = (0.202 A) 2 x 484Ω.

P100W = 19.74 W.

Confirmed power dissipated P80W > P100W i.e. Light bulb 1 (80W) is higher in power dissipation than light bulb 2 (100W). The 80W bulb is brighter than 100W light bulb when connected in series.

You may likewise find the voltage decrease throughout each light bulb and afterwards find the power dissipation by P = V x I as follows to confirm the instance.

V = I x R or I = V/R or R = V/I ... (Standard Ohm's Legislation).

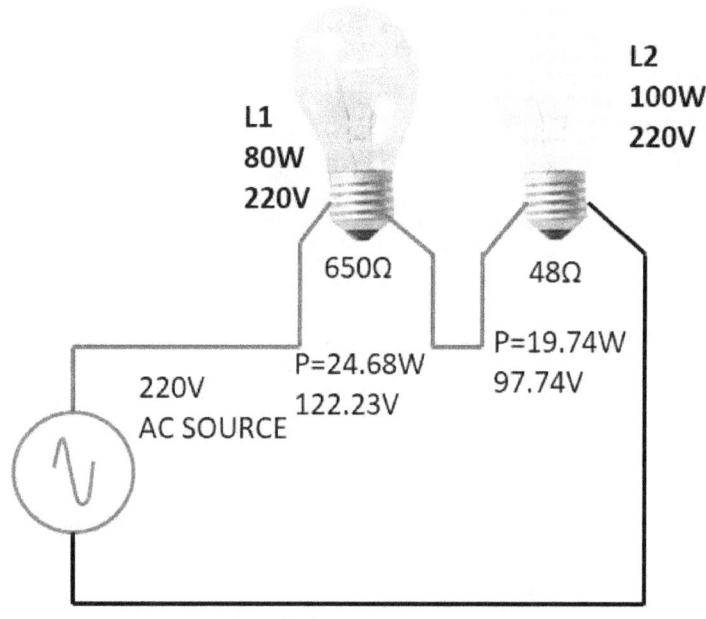

For Light Bulb 1 (80W).

$V80 = I \times R80 = 0.202 \times 605\Omega = 122.3$ V.

$V80 = 122.3$ V.

For Bulb 2 (100W).

$V100 = I \times R100 = 0.202 \times 484\Omega = 97.7$ V.

$V100 = 97.7$ V.

Currently,.

Power dissipated by Bulb 1 (80W).

$P = V280/R80$.

$P80W = 122.32$ V/ 605Ω.

$P80W = 24.7$ W.

Power dissipated by Bulb 2 (100W).

$P = V2100/R100$.

$P100W = 97.722$ V/ 484Ω.

$P100W = 19.74$ W.

Total Voltage in the series circuit.

$VT = V80 + V100 = 122.3 + 97.7 = 220$V.

Once again proved that 80W bulb is greater in power dissipation than 100W bulb when connected in series. 80W bulb will radiance brighter than 100W bulb when attached in series.

In Series Circuit, 80W Light bulb Glows Brighter due to High Power Dissipation rather than 100W Light bulb.

10.2. When Light bulbs are connected in Parallel.

Scores of light bulbs Wattage are various and attached in the parallel circuit:.

Currently we have the exact same two light bulbs each of 80W (Light Bulb 1) and also 100W (Light bulb 2) attached in parallel across the supply voltage of 220V ac. In that situation, the same will certainly occur i.e. the bulb with even more high as well as present power dissipation will certainly glow brighter than the various other one. This moment, 100W Bulb (2) will radiance better and bulb 1 of 80W will certainly dimmer. In other words, In parallel, both light bulbs have the very same voltage across them. The bulb with the lower resistance will perform more existing and consequently have a higher power dissipation and brightness. Baffled? as the instance has actually been turned around. Allow see the below computations and also examples to remove the confusion.

Power.

P = V x I or P = I2 R or P = V2/R.

Currently, the resistance of Bulb 1 (80W);.

We understand that voltages are the same in the parallel circuit as well as the rated voltage of light bulbs are 220V. i.e.

Voltage at Parallel Circuit: - VT = V1 = V2.....Vn.

Existing in parallel circuit: IT = I1 + I2 + I3 ... In.

R = V2/ P.

R80W = 2202/ 80W.

R80W = 605Ω.

And also, the resistance of Bulb 2 (100W);.

R = V2/ P.

R100W = 2202/ 100W.

R100W = 484Ω.

Now,.

Power raised by Bulb 1 (80W) and same voltage in a parallel circuit.

P = V2/R1.

P80W = (220V) 2/ 605Ω.

P80W = 80 W.

Power dissipated by Bulb 2 (100W).

P = V2/R2.

P100W = (220V) 2/ 484Ω.

P100W = 100 W.

Hence, showed P100W > P80W i.e. Light bulb 2 (100W) is higher in power dissipation than light bulb 1 (80W). Therefore, the 100W bulb is brighter than 80W light bulb when connected in parallel.

To confirm the above case, You may also discover the present for each light bulb and afterwards find the power dissipation by P = V x I as follows. We utilized the ranked voltage of the light bulb which is 220V.

I = P/ V.

For Light Bulb 1 (80W).

I80 = P80/ 220 = 80W/ 220 = 0.364 A.

I80 = 0.364 A.

For Bulb 2 (100W).

I100 = P100/ 220 = 100W/ 220 = 0.455 A.

I100 = 0.455 A.

Currently,.

Power dissipated by Bulb 1 (80W) as voltages are exact same in the parallel circuit.

P = I2R1.

P80W = 0.3642 A x 605Ω.

P80W = 80 W.

Power dissipated by Bulb 2 (100W).

P = I2R2.

P100W = 0.4552 A x 484Ω.

P100W = 100 W.

Overall Current in the parallel circuit.

IT = I1 + I2 = 0.364 + 0.455 = 0.818 A.

Again showed that 100W bulb is greater in power dissipation than the 80W bulb when linked in parallel. 100W bulb will radiance brighter than 80W bulb when linked in parallel.

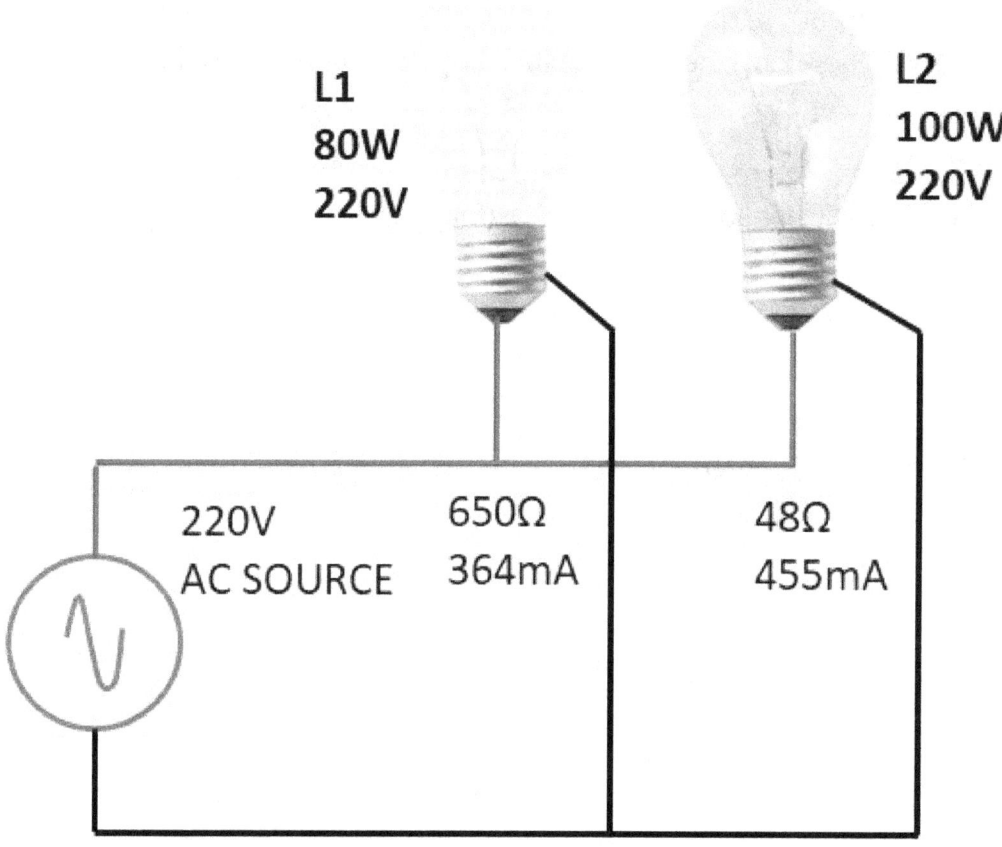

In Parallel Circuit, 100W Bulb Glows Better as a result of High Power Dissipation rather than 80W Light bulb.

10.3. Without Calculations & Examples.

The filament of the light bulb with a high rating is thicker than the reduced wattage. In our case, the filament of the 80W bulb is thinner than the 100W bulb.

In other words, 100 Watt light bulb has less resistance as well as 80 Watt light bulb has a high resistance.

10.4. When bulbs linked in Series.

We understand that present in a series circuit is same at each point indicate both bulbs obtaining the very same current and voltages are different. Obliviously, the voltage decline across greater resistance bulb (80W) will certainly be much more. So the 80W bulb will glow more vibrant as contrasted to 100W light bulb attached in series due to the fact that the same current is moving via both of the bulbs where the 80W bulb has more resistance because of reduced power level as the filament is thinner suggests it dissipates even more power ($P = V2/R$ where power is straight proportional to the voltage and inversely symmetrical to the resistance) and create higher warmth & light than the 100 W bulb.

10.5. When Light bulbs are linked in Parallel.

We additionally recognize that voltage in an parallel circuit coincides at each section which indicates both of the light bulbs have the very same voltage decline. Currently a load existing will certainly stream in the bulb which has much less resistance which is 100W bulb this moment which implies 100W light bulb dissipate even more power than 80W light bulb ($P = I2R$) where present and also resistance are straight symmetrical to the power. Therefore, 100W light bulb will glow brighter in a parallel circuit.

10.6. Just how to know if Light bulbs are attached in Series or Parallel?

A lot of the home electric circuit & installment are wired in series-parallel or parallel as opposed to series as parallel circuit has some benefits over a series circuit. So we may see that greater rated light bulb shines much more brilliantly as contrasted to reduced power level rated bulbs. Because case, 100W light bulb shines much more brilliantly than 60W or 80W light bulb.

Now, You should know that the light bulb with higher power score will radiance brighter when linked in parallel and the light bulb with much less power rating will radiance brighter in case of series electrical wiring as well as Vice versa.

11. Just how to Size as well as discover the Varieties of Ceiling Fans in a Space

Determine the Numbers of Fans and Correct Dimension of Ceiling Fans

Usually, the power electrical power of ceiling fans is 100 watts. For this reason, no greater than 6 fans can be connected to the last below circuit having 5 amperes current rating.

Fans, incandescent lamps and fluorescent as well as CFL lamps etc can be installed in the exact same circuit yet the overall current of devices ought to not boost than the ranked present of fuse/circuit breaker and also cable dimension made use of for the wiring installment.

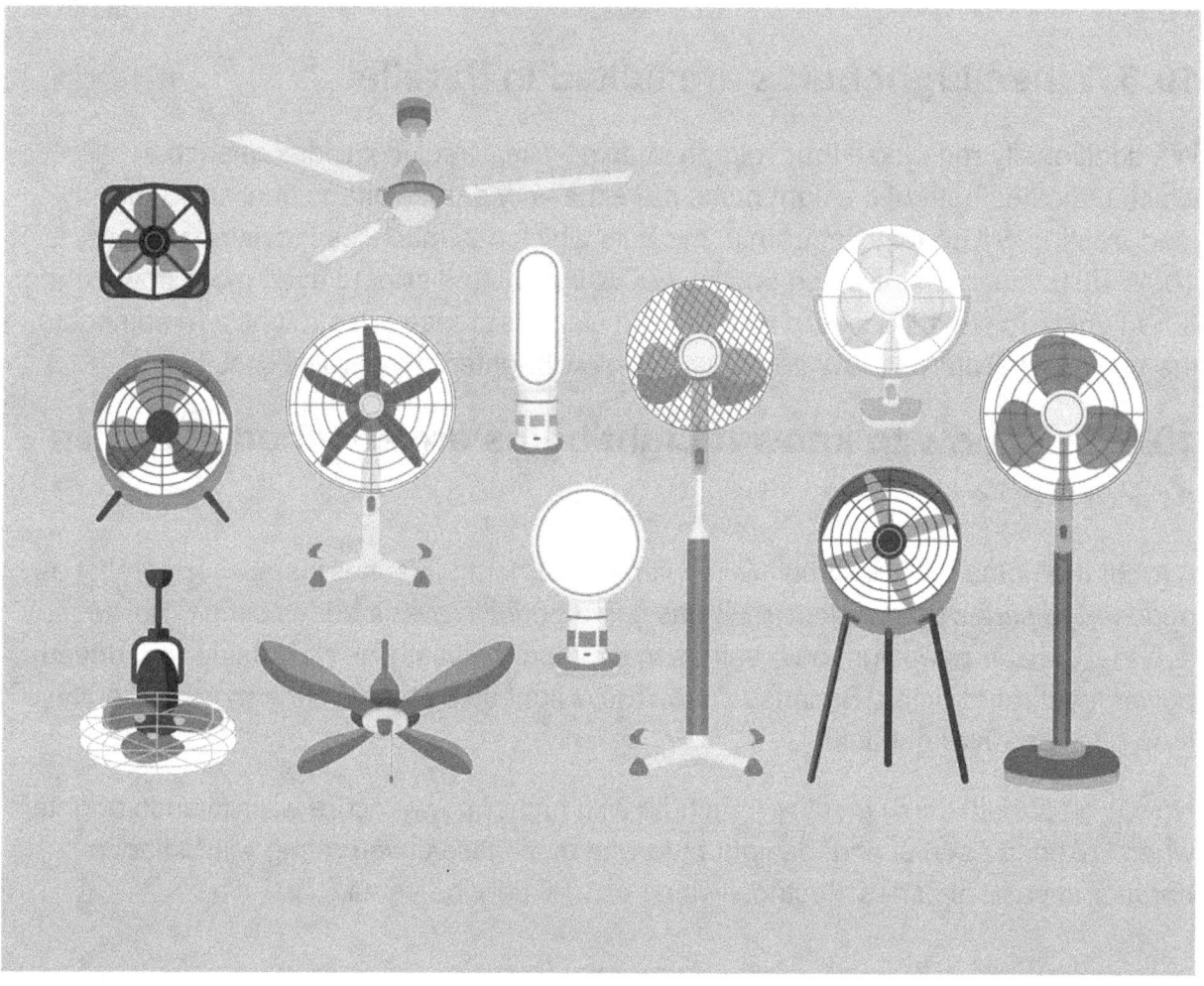

Each fan ought to be attached as well as control through different button and dimmer or speed regulatory authority (speed controller for variable speed). The minimal elevation of Rate regulator need to 1 1/2 meter.

To be on the risk-free side, limit size in between fan as well as ceiling ought to be 3 meters (10 feet) and also it should not be installed below 2.75 meters (9 feet). Ceiling fan need to be installed via hook or clamp and also 2 nut bolts of < 1/2" (12.7 mm) can be made use of to fix the rod in the hook/clamp.

Page intentionally left

Know about the
Backup supply-INVERTER

Residence Inverter, Exactly How to Pick and also acquire the most effective Inverter for Residence.

1. What is Inverter?

House inverter is the tool that powers the electric home appliances in the event of the power failing. Inverter as the name indicates initially converts A/C to DC for charge the battery and afterwards inverts DC to AC for powering the electrical supply. Various type of Inverters are currently available on the market and also of these the most reliable is the pure sine wave inverter which creates A/C comparable to the domestic power supply in wave type.

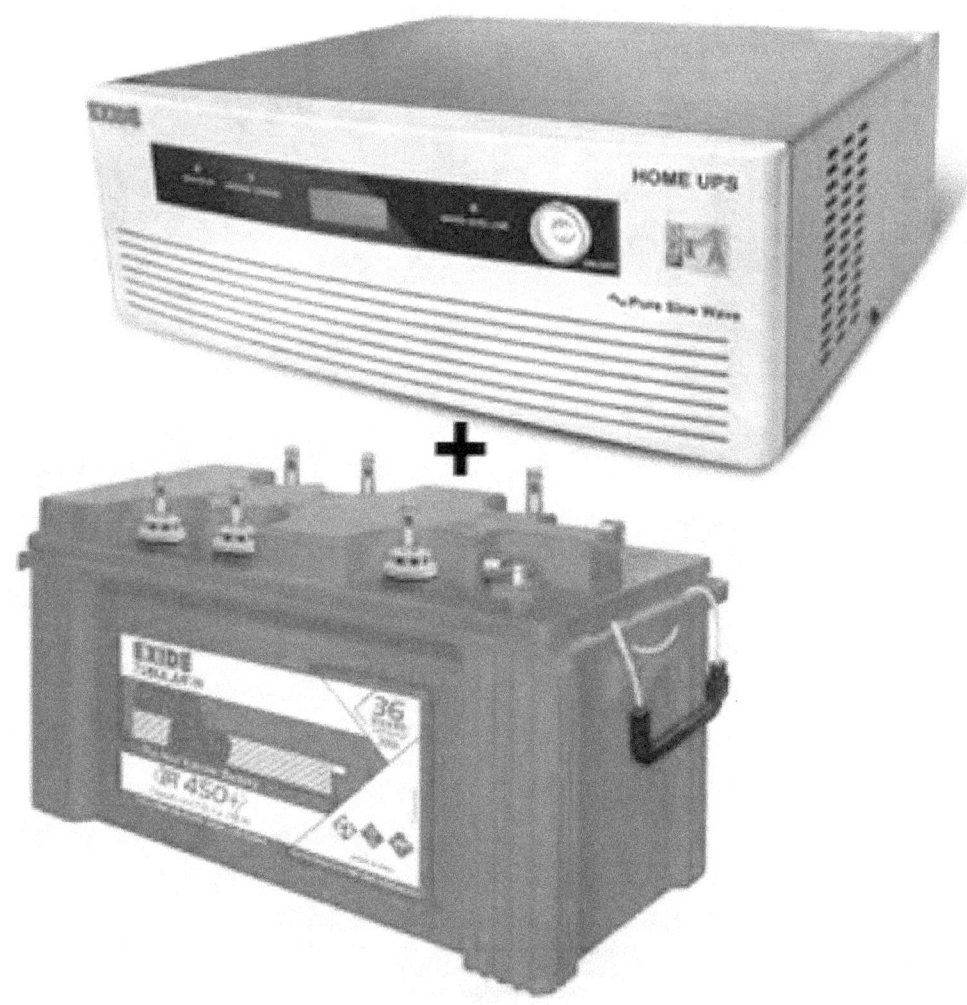

Square wave and also quasi sine wave inverters are usually low-cost types yet less effective than the pure sine wave inverter because some electrical appliances will certainly not work correctly in these inverters. Solar-powered Inverters are currently preferred to save energy yet its expense will certainly be extremely high considering that it requires a very large solar panel.

1.1. Residence Inverter

When the keys power is offered, the charger circuit charges the battery and the inverter area will certainly be stand by. The inverter section basically is composed of an oscillator and an inverter transformer.

The DC voltage from the battery is first converted into low volt ac by the DC-AC converter. The reduced vac is after that converted into 230 vac by a step-up transformer. Effectiveness of the power inverter depends upon the performance of the Oscillator as well as the Step-up transformer given that the frequency and voltage in the result relies on these sections of the Inverter. An easy house inverter circuit representation applied with transistors.

1.2. Residence inverter circuit representation

The foundation of the house auto power Inverter is the battery, which supplies DC for the inverter. The back-up time of the inverter relies on the ability of the battery. The inverter score is in regards to VA (Volt Ampere). 500 VA, 800 VA, 1000 VA etc are the usual residential inverters. Battery capability is stood for in Ah (Ampere hr). It is the ability of the battery to provide the quantity of existing in Ampere for one hour. For instance, a 100 Ah battery can supply 100 ampere present to the load for one hour.

1.3. Inverter Battery

Upkeep free batteries are utilized in Inverters since they need little interest. Upkeep complimentary inverter batteries are made using Apartment plate collection agencies and also calls for no water covering. Tubular batteries are much more reliable than Apartment plate types and they use Poly ester tubes full of Lead oxide instead of the Flat Lead plates. Battery deterioration due to plate rust is almost nil in Tubular batteries. They have more life span generally 5-6 years if properly kept.

1.4. Battery Capability and also Inverter efficiency:

The efficiency of the Inverter system mainly relies on the battery. It is an usual problem that "the inverter is not providing anticipated backup time". It is not the mistake of the Inverter, yet that of the battery. Battery capability is expressed as Ampere hour or Ah. 1

Ah amounts to 3600 Coulombs of power. Basically, 1 Ah battery offers 1 ampere existing in 1 hr. When the tons takes present, the battery discharges to ensure that the capacity of the battery minimizes as the discharge advances.

Normally a 100 Ah tubular battery can provide 5 Amps existing for 20 hours. The efficiency of the battery additionally depends on its cost/ discharge price. That is, the battery requires both charging and also releasing on a regular basis.

Appropriate charging of the battery is extremely crucial. If it goes down below 12 volts, the battery will not charge.

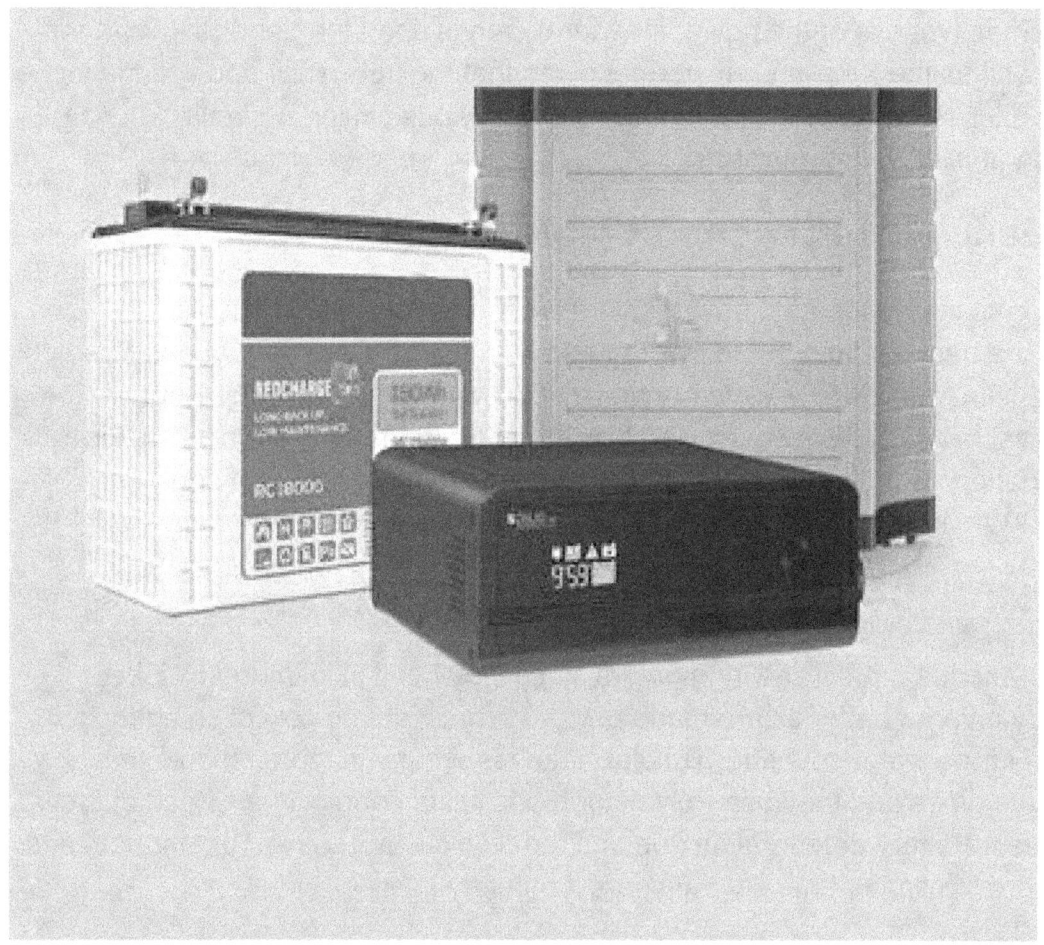

A completely charged Tubular battery shows 14.8 volts at the terminals. If this is listed below 12 volts even if charged for very long time, the battery is harmed and the back-up time will be considerably lowered. Throughout the very first few hrs of charging, the battery takes around 5-7 amperes present and then it will take only 500-700 milli ampere current in the subsequent hrs. A completely charged battery will not take any type of existing.

2. How to select the House inverter?

Before picking the House inverter, we have to calculate the back-up as well as the load time. To select the battery the adhering to formula can be made use of.

Complete load in watts/ Voltage of the battery x Back-up hrs called for.

As an example if the load is 400 watts as well as we need 3 hrs backup time, so the capability of the battery ought to be 400 watts/ 12 volt x 3 hrs =100 Ah.

A 100 Ah battery will not normally give a computed back-up time, because there will certainly be some power loss due to heating as well as in the supply lines. So the back-up time may be 2-2.5 hours relying on the problem of the battery. So it is far better to use next arrays such as 150 Ah or decrease the load's.

2.1. Picking Inverter based upon Load

Appropriate maintenance of the battery is necessary to boost its life. Constantly keep the battery and the Inverter in a dirt free ventilated area. The terminals of the Inverter must be firmly connected to the battery. Loosened link may lower current circulation as well as may likewise create triggering. The Tubular battery usually has water level indicators. Check the water level periodically.

If called for, top up with mineral complimentary battery water. Tubular battery requires water cover up only when in 6 months.

Inverter as the name suggests first converts ac to dc for charging the battery as well as after that inverts dc to ac for powering the electric devices. The backbone of the home car power Inverter is the battery, which gives DC for the inverter. Maintenance cost-free inverter batteries are made using Apartment plate enthusiasts as well as needs no water topping. Battery deterioration due to plate rust is virtually zero in Tubular batteries. When the load takes current, the battery discharges so that the capability of the battery lowers as the discharge advances.

3. UPS

An uninterruptible power supply (UPS) provides a simple solution: it's a battery in a box with adequate capability to run gadgets plugged in via its ac electrical outlets for minutes to hrs, depending on your needs as well as the mix of equipment. This may allow you keep net service energetic throughout a prolonged power outage, offer you the 5 minutes essential for your desktop with a hard disk to prevent and perform an automated closure shed work (or in a worst situation, running disk fixing software application). In regards to entertainment, it could offer you adequate time to save your game after a power outage.

A UPS also doubles as a surge guard as well as aids your devices and uptime by short-lived droops in voltage and various other vagaries of electrical power networks, a few of which have the possible to harm computer power supplies. For from concerning $80 to $200 for most systems, a UPS can provide an impressive amount of peace of mind combined with additional uptime and less loss.

3.1. Uninterruptible is the key word

The UPS raised in a period when electronics were breakable and drives were quickly shaken off kilter. They were designed to give constant-- or "uninterruptible"-- power to avoid a host of a problems. They were first located in web server racks and used with network devices until the cost and layout dropped to make them functional with house and small-office tools.

Any type of gadget you owned that unexpectedly lost power and also had a hard disk inside it may wind up with a corrupted directory site or even physical damage from a drive head striking another part of the device. Other tools that loaded its firmware off chips and also ran utilizing unstable storage can additionally wind up shedding useful caches of information as well as require a long time to re-assemble it.

3.2. 800AVR UPS

Hard drives advanced to much better take care of power failures (as well as acceleration in laptop computers), and all mobile devices and a lot of new computer systems transferred to movement-free solid state drives (SSDs) that don't have inner pins and also read/write heads. Installed devices-- from routers and modems to clever devices and also DVRs-- ended up being more resilient and quicker at booting. The majority of devices sold today have an SSD or flash memory or cards.

It's still feasible if your battery-free computer suddenly loses power that it may be left in a state that leaves a record corrupted, loses a spread sheet's most recent state, or takes place at such an inopportune moment you have to recuperate your drive or re-install the operating system. Preventing those possibilities, especially if you regularly run into small power issues in the house, can conserve you at the very least the moment of re-creating lost job as well as potentially the price of drive-rebuilding software application, even if your equipment stays intact.

A more typical trouble can emerge from networking tools that has small power demands. Losing power implies shedding accessibility to the internet, also when your dsl, fiber, or cable line stays powered or active from the ISP's physical plant or an area interconnection point, rather than a transformer on your building or block. A UPS can maintain your network up and running while the power business recovers the juice, even if that takes hrs.

The UPS's battery kicks in when power cuts out. It supplies anticipated quantities over all linked devices up until the battery's power is exhausted. A contemporary UPS can likewise signal to a computer system a variety of variables, consisting of staying time or activate a shutdown with integrated software application (similar to Power Saver in macOS) or mounted software application.

Page intentionally left

Know about the
Backup supply-SOLAR

Availability of Solar Panels to produce electrical energy in the houses has been there in all over the world now. Still, for a long time, it was not financially viable for the majority to take advantage. Yet as a result of a drop in photovoltaic panel rates in the last few years combined with considerable policy-level applications by Federal government of countries, currently, in 2020, Solar Panels for home looks fairly profitable.

Solar energy, a free renewable resource readily available at their disposal, lower their power expenses and increase eco-friendly impact effectively.

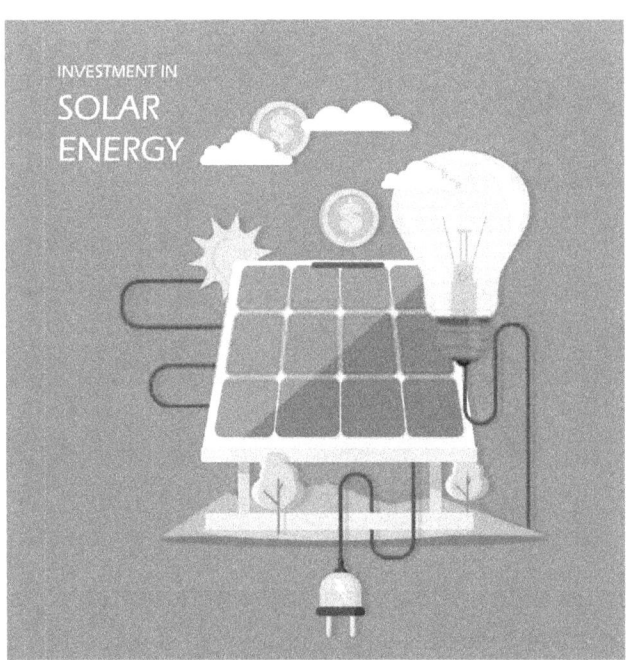

1. Perfect Solar Panel System Type for your house

There are two kinds of Solar Panels Solutions:

1.1. Off-Grid Photovoltaic Panel System:

This sort of system is ideal when you have normal power cuts in your house, and also you are seeking a power backup solution. This system consists of a Solar Panel, MPPT Charge Controller, Inverter and also a battery financial institution. The function of a Photovoltaic panel is to convert light into power (or electrical energy). The charged Controller ensures that the correct amount of electrical present is created and passed on to the battery. This stops any kind of damage to the batteries. The batteries are the storage tank that store all the power or electricity generated. The inverter resembles a vehicle engine, which assists run the appliances by taking electrical power kept in the batteries.

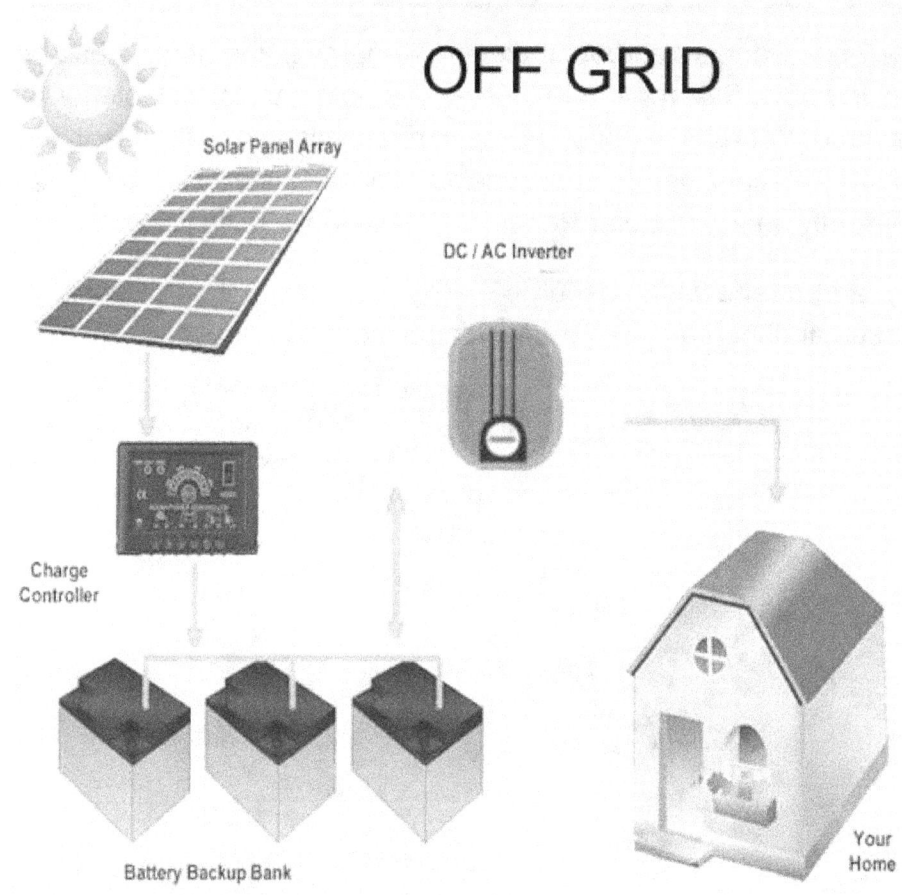

As this system features a battery, it can be used whenever there is no electrical energy. You can likewise obtain solar hybrid inverters that can charge the batteries when sunlight exists and also charge them from the grid when the sunlight is insufficient.

As this system involves batteries as well as storage, this system is pricey as well as likewise calls for regular upkeep of batteries. The batteries need a modification every 4-6 years relying on exactly how well they are preserved.

1.2. Grid Connected Photovoltaic Panel System:

If your objective of implementing solar is to decrease your electrical power expense, after that a grid-connected system is an ideal system. In a grid-connected system, there are no batteries. The power created is used in your home. If extra is produced, then it is marketed to your electrical energy circulation firm. If your power usage is greater than the production from the solar panels, after that it is compensated by the grid. This system has an inverter and also a net-meter.

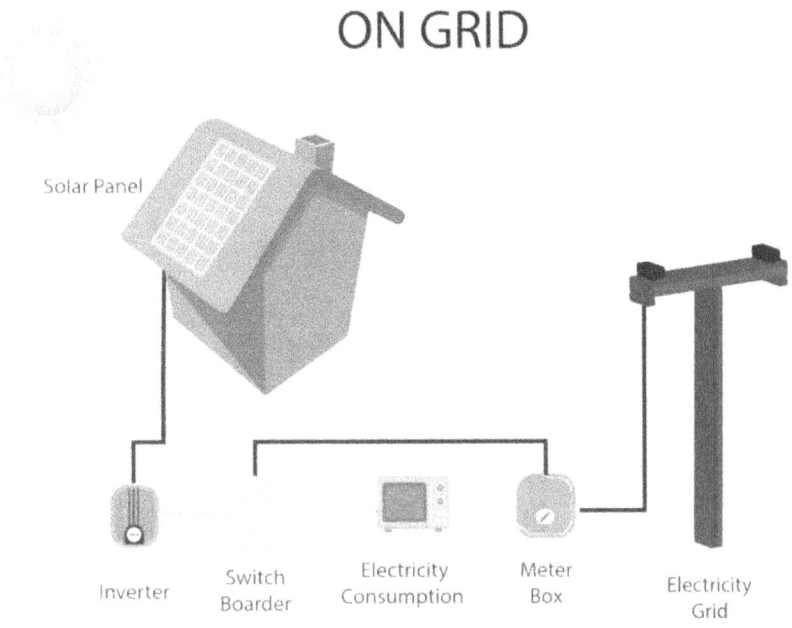

ON GRID

This system is bad if you have regular power cuts. That is because it decreases as the grid stops working. So if your system is creating power and the grid fails, after that the electricity produced is wasted.

2. Monocrystalline vs Polycrystalline.

There are 2 kinds of Solar PV Cells, Monocrystalline and Polycrystalline. The distinction in between both is that Monocrystalline is constructed from single silicon crystal. In contrast, Multi-crystalline PV is made up of numerous crystals. A monocrystalline is extra effective in transforming solar energy right into electricity per sq meter location than a multi-crystalline PV. Thus the area needed for the exact same amount of electrical power is less in monocrystalline PV panel. Therefore it is more expensive than a Polycrystalline PV.

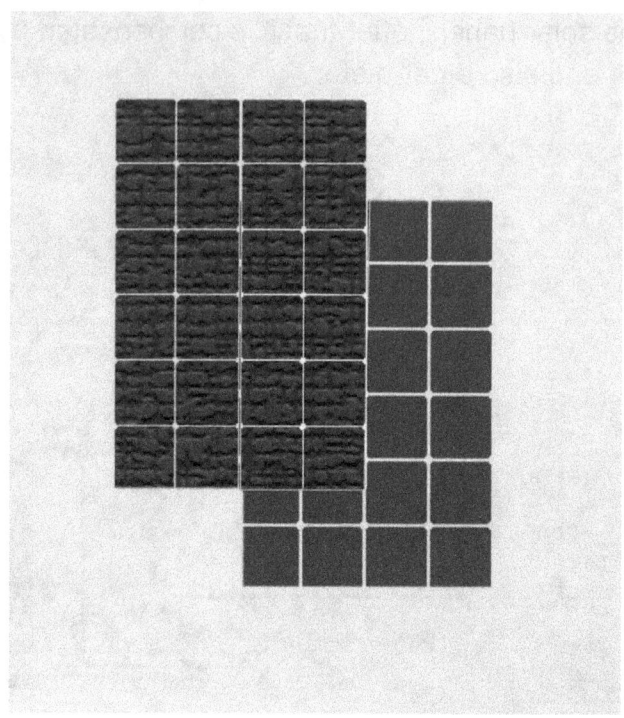

If you have much less area and also you have even more power needs, then go with Monocrystalline. Polycrystalline panels will certainly be great if you have enough room. To get a concept on room, read the next area on area requirements.

3. Space Needs of a Solar Panel System.

The area need of solar panels depends on the performance of the panels. A normal need is of regarding 80-100 sq ft for 1 kW system.

Please keep in mind that the space needed is not just any type of space, but it has to be shadow-free room where panels can be placed at the appropriate disposition. The proper disposition and also direction differ from city to city, and you ought to employ a Solar EPC (Engineer, Procure and Construct) company to aid you with it. Generally the excellent direction is South with an inclination of 15-18 levels. To get maximum out of your photovoltaic panels, you must likewise conduct a shadow evaluation of the location prior to installment. You need to see to it that:.

1. There is no outside body (like structures, trees) that casts a shadow on panels throughout the day.
2. Panels are positioned in such a way that they obtain optimal sunshine throughout the day at the appropriate angles.

There is software offered out there that can aid you do darkness evaluation. Here is a video revealing the effect of shadow on power generation:.

Numerous individuals ask us if solar panels can be installed up and down. People living in high rise structure ought to attempt to install a typical solar panel system.

4. Warranties and also Maintenance demands for a Solar PV system.

Solar Panels normally includes an efficiency service warranty of 25 years from the day of supply. The service warranties on inverter and batteries (in off-grid systems) differ from manufacturer to manufacturer.

The producers must provide a procedure, instruction and also upkeep manual in English as well as local language in addition to the system. Similar to all mechanical and electric system, solar panel system also needs normal upkeep. A reliable, long-lasting system is one that is maintained effectively and routinely. A solar panel system does not require a great deal of upkeep, however it is great to cleanse the system of dirt as well as bird droppings routinely to preserve its performance. If you pick a maintenance-free battery after that you need not stress over the battery, else the battery will certainly need routine upkeep.

5. Can your roofing hold the panels?

Commonly solar panels are rather light. Still, as soon as you have determined the place to install solar PV, you require to make certain that the roofing system needs to be able to take care of the solar PV system. If your roof is fairly high and also there is a whole lot of wind, you need to comprehend the configuration with your installer to make certain that the setup is secure sufficient and the panel must not dismount when there is a strong wind.

6. How to select the appropriate brand name for the Panel?

Now that you have actually figured out the best dimension of your solar PV, you need to start picking specific parts of your Solar PV system. When you call a company for Solar PV system setup, in most cases you won't find them to be producers of the elements of the system, they would simply be installers who help you pick the right parts. But after that it is necessary to be knowledgeable about the different brand names that are offered for those parts.

Consider this for an analogy: similar to vehicles have dealers as well as makers, where a dealership can have car dealership of one or numerous brand names of cars, a system installer for Solar PV will certainly have tie-ups with different brands of tools and they will certainly simply generate and also install the system in your property relying on what you select.

The first thing to pick is the panels. Much like while buying a vehicle, you get European (BMW, Audi, Volkswagen), Japanese (Honda, Toyota), Indian (Tata, Mahindra) and also Chinese products, in a similar way, in Solar Panels, you can get various brands from numerous countries. When it comes to Solar PV, the global market leaders are the brands of China.

Based on the data offered on Wikipedia, top worldwide brand names based on the capacity of installment, and also based upon our info from the sector are:.

1. First Solar (Malaysia).
2. Trina (China).
3. Canadian Solar (Canada).
4. Sharp (Japan).

If you are trying to find acquainted brands (names), after that you can look at the similarity:.

1. Mitsubishi.
2. Panasonic.
3. Bosch.

7. What is the distinction in between these brand names?

Top global brands will have more effective panels offering more outcome in less area. So, you can mount, say, a 1 kWp system in 90 sq ft with a top global brand name, whereas with a normal Indian brand you may obtain 1 kWp in 100 sq ft space. There might be some added benefits of reduced upkeep with international brand names. Yet after that you do not obtain aids if you opt for a global brand name.

Your panel supplier should give you a performance guarantee on the result of the solar panels. You should also get guarantees on the handiwork.

7.1. Just how to choose the best brand for Grid-Connected Inverter?

Just like panels, there are several brands of Inverters (from numerous locations) available in India. In the instance of inverters, your aid does not depend on the make of the inverter; you can acquire any type of inverter which fulfills MNRE requirements as well as obtain a subsidy.

Leading worldwide brand names of solar inverter (available in India) are:

1. SMA Solar.
2. ABB.
3. Schneider.
4. Fronius.
5. Huawei.

While selecting an inverter, please make certain that the efficiency of the inverter (to convert DC to AC) is high. International brands have power efficiency as high as 98%.

Which means, loss of just 2% and also almost all the DC energy that your PV panel creates obtains converted to beneficial AC power.

Inverters nowadays also have built-in software program solutions to keep an eye on and also take care of electricity generation from the photovoltaic panels. So, see to it that you get one that has this ability. If you go for an aid, after that the majority of the standard requirements (consisting of the software program) are covered in the requirements by MNRE.

Depending on the kind of your connection, you likewise have to ensure that it is a single-phase inverter or three-phase inverter.

A good warm sink with better warmth dissipation in addition to making use of electrolytic capacitors can make a great inverter, virtually a fit-and-forget type of system which you must search for calling for no maintenance for many years.

7.2. Just how to choose various other devices?

Other accessories required for a Solar Panel system are:.

1. AJB ('Array-Junction Boxes') for protection as there is no earthing.
2. AC and also DC circulation boxes-- for security as there is no earthing.
3. DC Cable.

While purchasing these, just make sure that you go with good brands like Havells, Finolex, Polycab or any leading brand that makes products based on BIS criteria.

Page intentionally left

Know about the

Backup supply-HOUSE GENERATOR

Genarator is also one of the great backup power source. Are you preparing to buy a generator for your home? We have you covered if so. In this chapter, we'll speak about different types of points as well as generators you should consider before buying a generator for your residence

1. Why do you require a generator in your house?

In this technology-driven age, we make use of a variety of devices, makers and gizmos in our homes to simplify our lives. From coffee devices, mixers, washing machines to laptops, A/C, TV, water cleansers, and so on, everything operates on power.

From coffee devices, mixers, washing equipments to laptop computers, AC, TV, water purifiers, and so on, everything runs on electricity.

We are a lot based on these points that in case of an electricity power outage, we feel completely helpless. No one wants their life to stop or to experience during power cuts. That is why it is an ideal option to invest in a generator.

There are myriads of generators that are readily available in the marketplace. Before purchasing a generator, you must take into consideration various variables so that you can get the best generator catering to your house needs and also spending plan.

1.1. Aspects to think about prior to acquiring a suitable generator for your house.

1. Spending plan

This is a prime factor while getting anything as no one prefers to go over their budget. So, prior to purchasing a generator, you should set your budget plan. You can find both high-end and also low-end products. Establishing your budget plan allows you to make an appropriate choice.

2. Electrical power needs

Each day, we make use of multitudes of electricity-driven things in our kitchen, bathroom as well as various other areas in our residences. So, it is essential to identify the power level requires or the power you would certainly require throughout outages. This can be figured out by including the wattage of all vital devices you wish to power. Usually, 3000-5000 watts is sufficient for a residence which is provided by mini-generators that are both low-cost as well as mobile.

3. Fuel utilized in the generator

Generators work on petrol, diesel, kerosene and also other fuels. So, you ought to choose the kind of generator that you would certainly desire in your home. Typically, portable generators that work on diesel appropriate for residence usage.

4. Vital and also special attributes

Like every product, a generator must likewise be finalized after checking out its attributes. Generally attributes like air conditioning, automatic transfer switch, automatic shut down, automated voltage policy as well as wheel package are preferable for an excellent generator. Considering that the generator is intended for residence usage, it must perform noise-free. So, it would be highly a good idea to pick a silent generator. The rate of silent generators for home is also quite cost effective

7. Select the kind of generator.

There are various kinds of generators based upon dimension, fuel-type and also power level. Each of these has its benefits and drawbacks as well as different cost variety. You can choose any type of one type catering to your requirements as well as budget.

8. Choose the size of the generator.

After making a decision wattage restrictions and also budget plan, you ought to select the dimension of the generator that is ideal according to your usage

Types of generators perfect for house usage.

2. Standby generators

These generators need to be set up outside a structure. They supply power in between 5000 to 15000 watts.

2.1. By size

1) Inverter or little generator: These are small sized generators that offer power between 2000 to 7000 watts. They are mobile generators ideal for outdoor camping and also residence use.
2) Medium generator: These supply power in between 8000 to 20000 watts.
3) Huge generator: They supply 20000 to 40000 watts power.

Generally, it is excellent to utilize a mini generator for house as it is affordable, portable and also sustains all the basic appliances.

2.2. By fuel type

1) Diesel generators: These generators run on diesel. They are good for long-lasting usage. Given that diesel is less costly so these are economical.
2) Gas generators: These generators run on gas which is cheaper than all gas types. They have a greater service life.
3) Kerosene generators: These are the earliest kinds of generators yet nowadays, it is hard to discover kerosene. Or else, they are the least expensive. They must be correctly ventilated as well as cooled down to stay clear of damage due to overheating

4) Petrol generators: These are quite effective in providing a huge quantity of power. Due to the expensive price of fuel, they are perfect just for small procedures.

The electrical power demand of a home or little shop is typically listed below 5000 watts. So, if you are seeking an economical choice, after that a standby or mobile generator is perfect for residence use. These generators can powering all the major devices utilized in a family.

Page intentionally left

How to Select your Home -Air Conditioner

Acquiring a brand-new air conditioning system is a big choice, not even if it's pricey, however since it determines whether you fit in the house or not. Everybody expects smooth and also effective cooling when they purchase a brand-new system, but regrettably, that's not constantly the case.

OVER FIFTY PERCENT of all brand-new HEATING AND COOLING systems are poorly mounted, which leads to decreased efficiency and reduced lifespans. Not only that, comfort problems can create as well, such as high humidity, air-borne pollutants, as well as inadequate air blood circulation (Power-Efficiency). One of the most important day of an a/c unit's life is the day of its installation, cooling is a requirement in a lot of houses. Also in position where the environment is fairly modest year-round, an A/C unit is vital for both the comfort and also health of a family. Along with maintaining an interior cool, cooling keeps the air circulating, helps reduce humidity, as well as can aid remove harmful airborne contaminants.

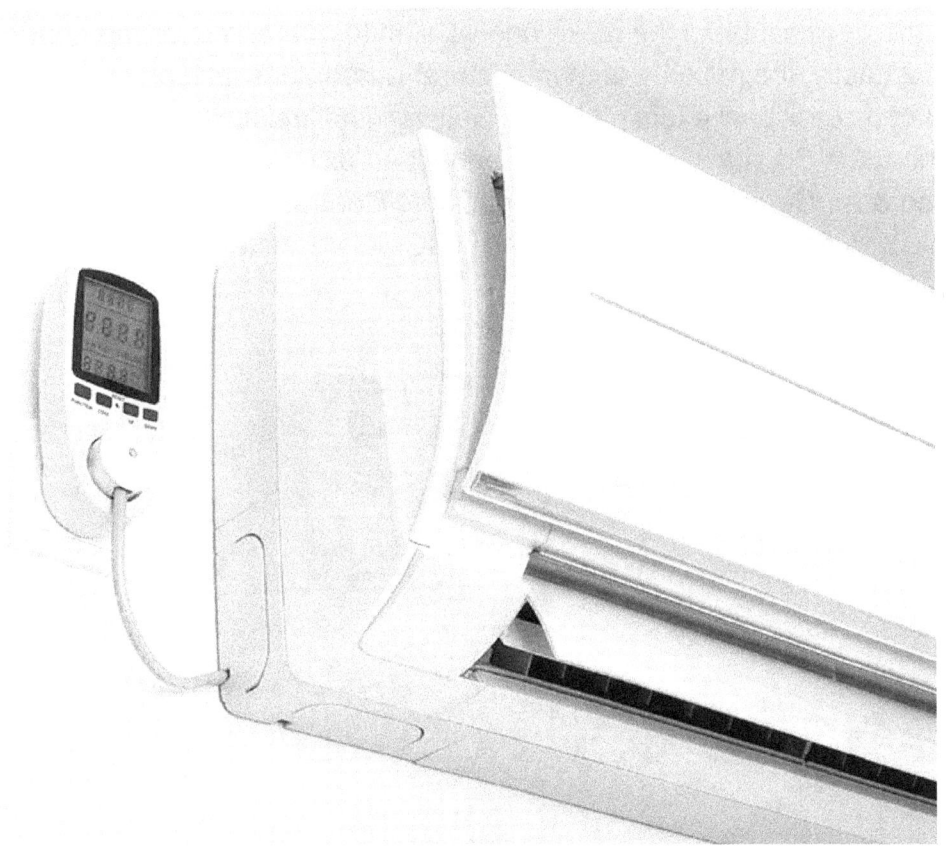

New units last generally 15-20 years, so it's a vital financial investment that ought to be completely researched before choosing. Even if your A/C is working, you might be able to conserve more money by replacing your unit as a result of the raised energy savings of brand-new devices. Talk with an A/C professional to identify if it makes more sense to replace rather than fixing if your air conditioning system is 10 years or older.

If you are considering getting a new air conditioning system, there are several things to consider. Whether you're changing an old unit or require to acquire a new residence, we've assembled some crucial tips for your search.

COOLING AND HEATING Customer's Overview:

What to Know Before Getting a New A/c Unit

1. Obtain a Professional Dealer

It is imperative that you investigate the most effective contractor/company for the task. Your air conditioning system's efficiency is established by its setup. A referral from a friend or member of the family is constantly the very best course, but regional home service referral websites available in BBB ranking too. Ensure you select 2 or 3 various contractors and also have each deal a thorough quote. If they are accountable for supplying the device, obtain the make and version to guarantee it's the right one for your needs. Lastly, the majority of manufacturer guarantees include certain terms, consisting of professional examination as well as installment, signed off by a qualified HVAC contractor. So, make certain you conduct an extensive vetting procedure. Utilize this Heating & Air Conditioning Installment Quote Contrast List by ENERGY-EFFICIENCY to screen your HEATING AND COOLING installers.

2. Online Purchase

The initial is complete where the service provider provides both the installation and also the system. You might decide for purchasing an unit at wholesale and employing a service provider for the installment. Numerous service providers will only set up COOLING AND HEATING units they have supplied themselves due to the fact that maker warranties go with the business and also your unit might be harmed, the incorrect size, do not have proper documentation, or incorrectly matched for your existing house.

3. A/C Dimension

Choosing the proper size system for your home is crucial. While your professional will aid you with this part of the formula, it's important to understand in advance why this is essential. If a unit is the incorrect size for the square video of your house, a number of issues will certainly result. , if the device is too small it won't be able to effectively cool your residence. If the system is as well big it will cycle on and off often-- wasting energy and also inflating your energy expense. Even if you are replacing an older design, your professional will certainly want to execute what is called a tons estimation.

4. Load Computation.

If you are getting a brand-new cooling device, you will certainly desire a tons estimation executed. A tons estimation assists to determine the suitable dimension of the A/C your home will certainly require. Your HEATING AND COOLING technician does this by matching your residence's thermal characteristics with cooling ability (in BTUs). A common estimation takes into consideration the size of your house, climate area, roofing product, directional alignment, variety of devices and other important information. Examine if your A/C specialist performs power audits too. You can frequently combine power audits as well as load calculations with each other for a price cut.

5. SEER Rating

A cooling system's effectiveness is gauged by the Seasonal Energy Performance Ratio (SEER). In 2015, the UNITED STATE Department of Power (DOE) elevated their minimal SEER demand from 13 to 14. Bear in mind that a score of 14 to 22 is thought about to be an energy-efficient category. So look for a version that is rated at the very least a 14, however look for one that is even greater on the efficiency scale. Discover more HVAC terms right here.

6. High-Efficiency Versions

A high-efficiency A/C unit (rated 14+) will certainly cost a little bit a lot more, but it will eventually save power (and money) in the future. Search for the ENERGY STAR tag when choosing a new ac system as it identifies one of the most efficient systems on the marketplace.

Yearly Upkeep-- After your brand-new device is installed you will certainly likewise want to talk to your service provider concerning an upkeep plan. Many professional home solution firms have bargains on annual A/C upkeep.

Page intentionally left

How to Select your Home -Refrigerator.

Selecting the Best Refrigerator Design for Your Cooking area

1. Which fridge is best for your kitchen area?

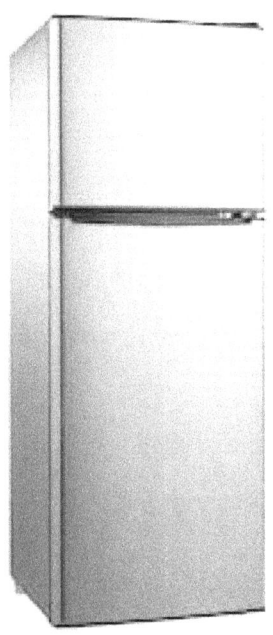

There are several ranges of refrigerator designs, as well as it can be really hard to determine which sort of refrigerator will work best for your cooking area and also household. While installment constraints as well as capacity demands usually dictate how big a design you should buy, how that fridge is created (style) likewise issues.

The design of fridge can impede or improve your kitchen workflow, creating a lot more efficient (or otherwise) dish prep which conveniently translates right into saving time and energy. For a busy household, this can make a significant difference in your day. Perhaps even in just how your day starts.

Refrigerator design doesn't just relate to shiny coatings, bells, and whistles (how it looks), yet how the general style of this device operates in your cooking area. These will certainly assist you to consider alternatives as well as eliminate the styles that are bad choices, to limit the best design for your certain needs. After that, you'll prepare to start buying a refrigerator.

2. All Fridge

- This is typically the most affordable version
- Looks similar to an upright fridge freezer
- Not too many frills; the very least attributes on this design
- You can better organize all your cooled foods
- Easy access to top half for things you utilize usually

- Offers the biggest cooled ability (no fridge freezer compartment).
- Best for houses that likewise have a fridge freezer which is conveniently situated near the kitchen area.
- Excellent option for an extra or 2nd refrigerator.
- These are typically hand-operated defrost. Confirm the sort of cleansing before purchasing.
- Less capacity option for this style of fridge; more small designs.

3. Single-Door Refrigerator with Freezer Capability.

- This style has a freezer area, however there's just one fridge door, not a split door.
- Affordable to buy.
- Less attributes, less capability, or less choices of outside coating.
- Less energy effective. Every time the door is opened warm air can cause the fridge freezer temperature level to climb as well as power is required to cool off once again.
- Freezer compartments are insufficient as well as usually little.
- Single door swing; enable adequate area for door opening.
- Generally available in smaller capacities.
- Excellent choice for the budget plan minded, tiny spaces, or as an additional system.
- Best if there's likewise a freezer in the home.

4. Top (Freezer) Mount Refrigerator.

- Most prominent model; perfect for smaller cooking areas.
- This style has the freezer on the leading with a different split door.
- Traditional and also affordable design.
- Fridge freezer temperature remains a lot more consistent with a separate door.
- Freezer compartment sizes vary per model.
- Great option of attributes, finishes, as well as abilities.

- Allow room for full door swing.
- Best selection for those who do not have space for a French door or side-by-side version.

5. Bottom (Freezer) Mount Refrigerator.

- Fridge freezer area is on all-time low with a different door or cabinet.
- More costly than a top install fridge.
- Freezer compartment capacity varies.
- Good selection of surfaces, attributes, racking.
- Excellent variety of capacities and versions readily available.
- Various fridge freezer configurations; shelving or pull-out basket with a door or take out freezer drawer.
- Usually has a larger chilled area.
- Chilled things used usually go to eye level.
- More flexing for some to easily access the bottom fridge freezer; more difficult to arrange materials.
- Before buying, evaluate the fridge freezer access for comfort.

6. Side by Side Refrigerator.

- A lot more costly than traditional top or lower place.
- The freezer is normally on the left side; refrigerated on the right.
- Versions range from 22 cubic feet and also up.
- Abilities vary depending on width, elevation, and also deepness.
- Offered with or without water as well as ice makers/dispensers.
- Optional water or ice dispensers call for pipes set-up expenses.
- Less refrigerated ability, yet much more freezer capability than a top place.
- Fridge freezer convenience for organizing as well as fetching foods.
- Odd-sized trays or things may need creative positioning because of minimized food storage space size.

- Allow space for (short) door openings on each side; much less door swing than lower or top place fridges.
- Good range of attributes, abilities, and also coatings.
- Good selection for small cooking areas as a result of the shorter door swing.
- Perfect for those who wish to get rid of the need for a different fridge freezer.
- Easier fridge freezer gain access to for some, contrasted to the bottom install version.

7. French Door Models.

- Most pricey, yet the majority of hassle-free specifically for food trays.
- Split doors with the biggest chilled capacity.
- Freezer gets on all-time low; setup can vary.
- Excellent selection of attributes, flexible racking, containers.
- Some versions have 4 doors or extra delicatessens cabinet.
- Not offered in under 20 cubic foot refrigerator designs.
- Optional water/ice dispensers-- calls for plumbing set-up.
- Best for storing odd and also cumbersome things.
- Finest use of cooled space.
- Popular systems for storage comfort, trendy appearances, option of finishes and also brief door swing.

8. Actions to Making a Refrigerator Style Decision.

Beginning by taking dimensions of the area where the refrigerator will certainly be installed, including the permitted elevation, if you have overhead cupboards. Enable sufficient width for door swing(s).

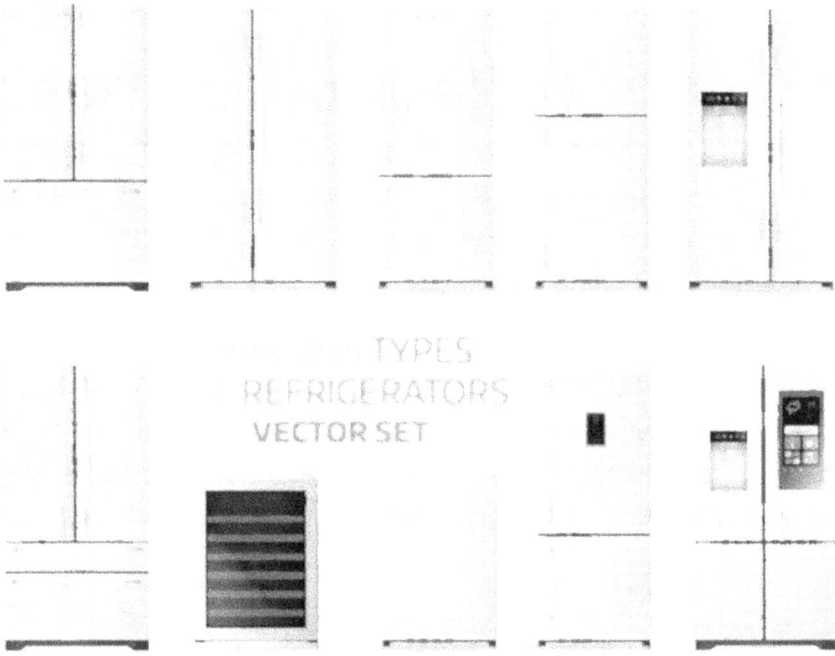

Refrigerator versions can be counter/cabinet or full deepness which impacts space and ability. Find out more about fridge depth choices to more limit your design options. If you 'd like a model with onboard water and also ice, understand that this feature will minimize the total ability. It will likewise require a plumbing connection.

Next off, do you desire a fridge freezer compartment as well as where would certainly you like it to be: on the side (side-by-side), under (bottom place) or on the top (top place). Possibly, you already have a standalone fridge freezer close-by and you 'd like an all fridge design. That will certainly conserve money, however you'll have less model range to choose from.

Each design of refrigerator has its comfort level. Attributes, shelving, and moisture controls vary per model as well as naturally, will certainly affect the bottom line. You can aid to maintain prices down by getting just food storage space includes that matter to you if you have actually nailed down your selections to a French door version. Prices for this prominent fridge style can be substantial, so select wisely.

Estimating Device and also Residence Electronic Energy Use

Our device as well as electronic power use calculator allows you to estimate your annual power usage as well as cost to run specific items. The wattage values are provided for sample purpose only.

Identifying how much electrical power your appliances as well as house electronic devices make use of can aid you understand just how much cash you are spending to use them. Utilize the information listed below to approximate just how much electricity a home appliance is using as well as how much the electrical power sets you back so you can make a decision whether to invest in an extra energy-efficient appliance.

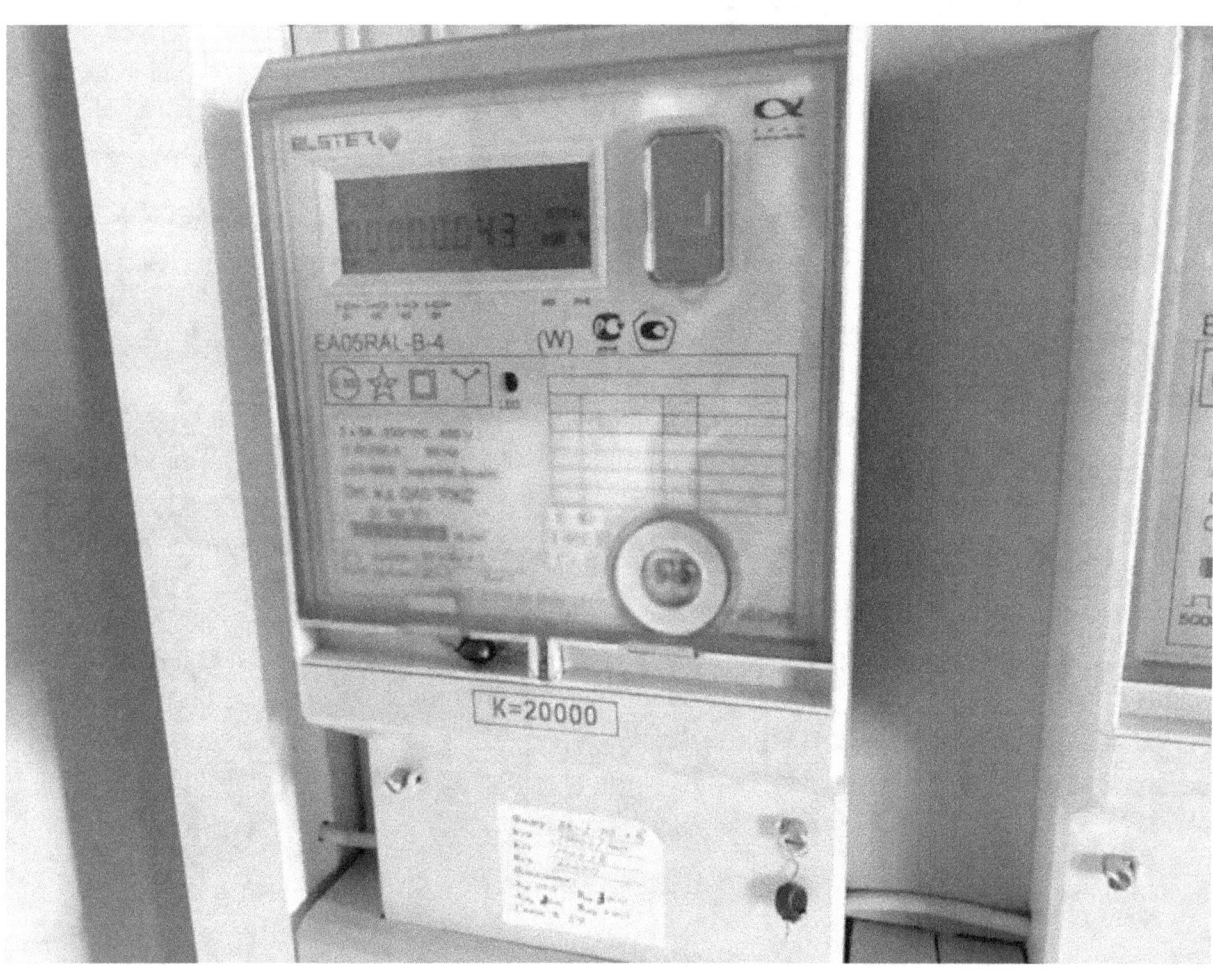

There are numerous ways to estimate just how much power your appliances and also residence electronic devices utilize:

1. Assessing the Power Overview tag. The label supplies an estimate of the typical energy intake and also expense to operate the details version of the appliance you are using Keep in mind that all not all devices or house electronics are called for to have a Power Overview.
2. Utilizing an electricity usage monitor to obtain readings of how much electrical power a device is utilizing.
3. Determining yearly power intake and also prices utilizing the formulas offered below
4. Installing a whole house energy surveillance system.

1. Energy meters

Electrical energy meters screens are easy to use and also can determine the electrical power use of any type of device that works on 120 volts. (Yet it can not be utilized with large home appliances that make use of 220 volts, such as electric garments dryers, central air conditioning conditioners, or hot water heater.)

Screens are useful for finding the quantity of kWh used over any type of time period for tools that don't run frequently, like fridges. Some monitors will let you enter the quantity your energy costs per kilowatt-hour and also give a quote how much it set you back to run the device considering that it was linked into the screen.

These phantom loads take place in the majority of appliances that utilize electrical energy, such as televisions, stereos, computer systems, and also kitchen area appliances. These loads can be prevented by unplugging the home appliance or making use of a power strip and also utilizing the button on the power strip to reduce all power to the device.

2. Calculating Annual Electrical Power Intake and also Prices

Follow these actions for locating the annual energy intake of a product, along with the cost to run it.

1. Estimate the number of hours each day a home appliance runs. There are 2 methods to do this:

2.1. Rough estimate

If you learn about just how much you use a device each day, you can approximately approximate the number of hrs it runs. For instance, if you understand you typically view regarding 4 hours of tv each day, you can use that number. If you recognize you run your whole residence fan 4 hours every evening before shutting it off, you can make use of that number. To approximate the number of hours that a fridge really runs at its optimum electrical power, separate the total time the fridge is connected in by three. Fridges, although transformed "on" regularly, really cycle on and off as required to keep interior temperature levels.

2.2. Keep a log

It may be useful for you to maintain an use log for some home appliances. You could videotape the cooking time each time you utilize your microwave, job on your computer, view your television, or leave a light on in an area or outdoors.

3. Locate the wattage of the product. There are 3 ways to locate the power level a device utilizes:

3.1. Stamped on the home appliance

Several home appliances have a range of setups, so the actual quantity of power a home appliance may take in depends on the setting being made use of. A radio established at high volume uses even more power than one set at low quantity.

3.2. Increase the appliance ampere usage by the appliance voltage usage

If the electrical power is not provided on the appliance, you can still estimate it by finding the electric current draw (in amperes) as well as multiplying that by the voltage used by the appliance. Most appliances in the USA utilize 120 volts. Bigger appliances, such as clothes dryers and also electric cooktops, make use of 240 volts. The amperes may be stamped on the system instead of the power level, or listed in the owner's handbook or spec sheet.

3.2.1. Use online resources to locate normal electrical powers or the electrical power of particular products you are thinking about purchasing. The complying with web links are good options:

The Home Power Saver provides a listing of appliances with their estimated wattage and also their yearly energy usage, in addition to various other attributes (consisting of yearly energy use, based upon "normal" usage patterns. Continue utilizing the formulas below if you want to discover energy usage based on your own use patterns).

In some instances, you can use the supplied info to do your very own quotes utilizing the formulas below. The info might also aid you compare your existing home appliances with extra effective models, so you understand possible cost savings from upgrading to a much more reliable home appliance.

3.2.2. Discover the day-to-day energy usage utilizing the following formula:

(Electrical Power × Hrs Utilized Daily) ÷ 1000 = Daily Kilowatt.-hour (kWh) intake

3.2.3. Discover the yearly power usage utilizing the adhering to formula:

Daily kWh consumption × variety of days utilized annually = yearly energy intake

3.2.4. Locate the annual cost to run the device using the complying with formula:

Yearly power consumption × utility rate per kWh = yearly cost to run device

I. Instances:

I. Adhering to the actions over, locate the yearly expense to run an electrical kettle.

1. Price quote of time utilized: The kettle is used several times per day, for regarding 1 overall hour.

2. Electrical power: The wattage gets on the tag as well as is provided at 1500 W.

3. Daily power intake:

(1500 W × 1) ÷ 1,000 = 1.5 Kilowatt hour

4. Annual power usage: The pot is utilized nearly daily of the year.

1.5 kWh × 365 = 547.5 kWh

5. Yearly cost: The energy price is 11 cents per kWh.

547.5 kWh × $0.11/ kWh = $60.23/ year.

II. Complying with the actions over, find the yearly expense to run a paper shredder.

1. Estimate of time made use of: The shredder is utilized for around 15 mins daily (0.25 hour).

2. Wattage: The power level is not listed on the label, however the electric present draw is listed at 3 amperes.

$120V \times 3A = 360W$.

3. Daily power intake:.

$360 W \times .25 \div 1000 = 0.09$ kWh.

4. Annual energy usage: The shredder is made use of regarding once per week (52 days each year).

0.09 kWh $\times 52 = 4.68$ kWh.

5. Yearly price to operate: The utility rate is 11 cents per kWh.

4.68 kWh $\times \$0.11/$ kWh $= \$0.51/$ year.

4. Whole-House Power Checking Systems.

If you desire more in-depth data on your home's power usage (along with the capability to determine the energy use 240-volt devices), you might think about installing a whole-house power surveillance system. The attributes of these systems vary, and the price as well as intricacy depends upon the variety of circuits you intend to keep track of, the degree of information of the information, as well as the functions offered.

Along with the present loads on the energy intake of your appliances, these monitors provide you comprehend where as well as when you use one of the most power, allowing you to establish methods to decrease your power use as well as costs. Again it depends where you are living.

Leading suggestions to reduce your Power bill

Heating devices can have a significant impact on your energy costs, particularly when they're not correctly preserved. Picture/ Getty.

As winter effectively sets in, you have actually likely found yourself with an expanding power costs.

Last year, the ordinary Kiwi house spent $2101 on power, concerning $160 each month.

So with shorter days needing lights on earlier as well as colder evenings luring us to switch on heating systems and also electric blankets, exactly how can you keep your power use down?

1. Change to LED lights.

Much shorter winter months days mean we're turning on lights earlier and utilizing even more power. You can alleviate your electricity use by altering light bulbs to LEDs. They're a greener option and are likewise a lot more energy efficient, using up to 80 percent less energy as well as long-term 25 times longer than incandescent lights.

LEDs are readily available to match in a range of sizes and shapes, from down lights to floodlights. Changing your light fittings to dimmable LEDs not just saves you power, yet their lengthy life produces easy maintenance.

Vanessa advises making use of cozy white in living locations to give them a cozy feel, as well as amazing white in various other areas like washrooms and also kitchens where a crisper light is needed.

Savings.

1. Changing to energy-efficient illumination could halve your home lights prices.

2. According to Energy.gov, changing your home's 5 most often utilized light bulbs with models that have earned the power star can conserve up to $75 a year.

2. Turn off at the wall surface.

When we're not utilizing them, it's simple to do but we commonly forget to turn off our light buttons and also devices at the wall.

Several house appliances, especially entertainment systems and technology gadgets, still chew out power even when they get on standby setting. You can save yourself some money by unplugging every one of your appliances or using a multi-plug board so all devices can be shut off correctly at the same time when you're do with them - consisting of the children' Xbox.

Savings.

According to Energy Wise appliances left on standby can cost you greater than $100 a year on your power bills.

3. Get your timing.

Homes lose a great deal of power via unneeded heating. A programmable timer can assist maintain losses to a minimum by letting the temperature level in your house stay at the best level to match you.

You can use a timer to set the air conditioning (heating) to "off" during the day, when the house is empty, and program it to turn on soon prior to you schedule home. This suggests you're coming home to a warm, cosy residence. You can likewise do the very same with your towel rail.

Savings.

According to Power Wise if you have your towel rail on for four hrs a day instead of constantly you could save $130 a year.

4. Display your usage.

If you do not understand where you're using the most electricity, cutting down your power consumption can be hard.

There is a series of products available that help you monitor your electrical energy consumption around the clock by means of an online app.

A PDL product called Eco mind permits you to keep track of your power usage, evaluate your usage and also discover exactly how to save power. It's also a terrific way to teach children regarding conserving power.

5. Maintain your home appliances.

With winter season's chilly chill, it's hard to withstand the convenience of a natural gas heating unit or reverse cycle air-conditioner. But the reality is these appliances can have a significant impact on your energy bill, specifically when they're not appropriately kept.

Performance is impaired leading to greater costs when filters have a build-up of dirt. Ensure your family home heating is constantly running at its optimum by cleaning up the filter as soon as a week. Much better still, placed a pointer in your phone so you can establish as well as neglect.

It's likewise an excellent concept to routinely check your stove, fridge as well as fridge freezer seals to make sure you're not making use of additional power due to a leak.

Preserving your power outlets as well as switches will additionally conserve you cash. Due to the fact that they consist of real-time wires, harmed home electrical can not just make use of excess power however are a potential safety and security worry. Not only will this aid you reduce your energy costs yet it will prolong the life time of your appliances.

6. Take faster showers.

When it's wintry outside, a long, hot shower can be challenging to stand up to. Research study programs hot water accounts for 21 per cent of all energy utilized in the house.

Maintaining shower sizes to a minimum, preferably 2 minutes, is a very easy as well as quick means to downsize your bill.

Savings.

According to Energy Wise a 15 min shower expenses around $1, a 5 min shower around 33c. That indicates a family of 4 could be conserving around $18 a week, or $900 a year, just by taking much shorter showers.

Page intentionally left

Know about Energy Efficiency

- STAR RATING

Concerning Standards & Identifying Program

Energy efficiency STAR Rating are applicable in Ac system (Repaired Rate), Ceiling Fan, Colour Television, Computer System, Direct Cool Fridge, Distribution Transformer, Domestic Gas Oven, Frost Free Refrigerator, General Objective Industrial Motor, Monoset Pump, Openwell Submersible Pump Establish, Fixed Type Water Heater, Submersible Pump Establish, Tfl, Washing Machine (Semi/Top Load/Front Load), Ballast, Solid State Inverter, Office Automation Products, Diesel Engine Driven Monosetpumps For Agricultural Purposes, Diesel Generator Establish, Led Lamps, Room Ac System (Variable Speed), Chillers, Agricultural Pumpset, Microwave Oven, Deep Freezers, Light Commercial Air Conditioning Fixed Rate, Light Commercial Air Conditioning Variable Rate.

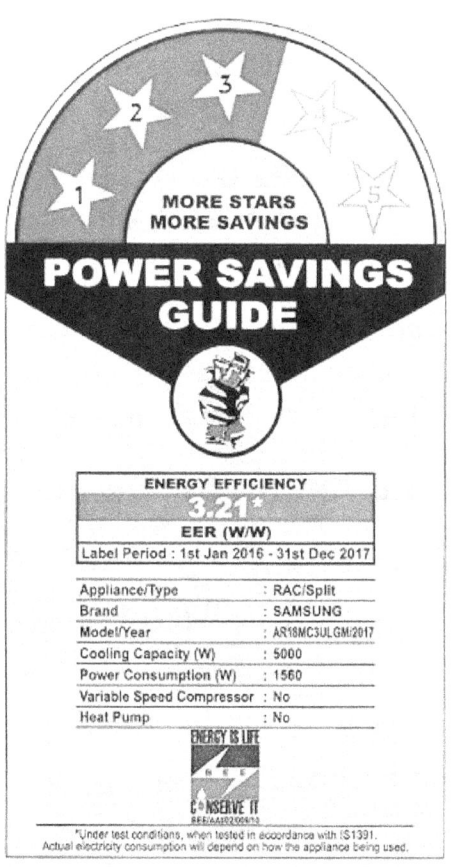

Whenever you're looking for devices such as fridges, a/c, or hot springs, you might have observed star rating sticker labels on the appliances. These are called SEER/ BEE star tags and also they show how much electrical energy the device consumes in a year. Many countries are following their own standards to declare the star rating.

Each home appliance gets between one as well as five stars, with five stars suggesting that it's exceptionally efficient and also is most likely to keep your power expenses in check.

Right here's whatever you require to find out about star tags, but this concept is more or less similar to the all countries.

1. How are star rankings calculated?

SEER states that obtains all of its information straight from the producers. Whenever makers in particular item classifications such as a/c unit introduce a new product, they get it examined in a laboratory recognized by Government.

Checking these home appliances is not a long process. It might take simply a day to obtain a design examined in an approved lab. Great deal of values are theorized. For e.g. in situation of air conditioners, the power usage and also performance are evaluated at 100 percent as well as 50 percent capabilities. And then the worth is extrapolated to running conditions in various seasons.

Yes, it's true that the company doesn't test each home appliance prior to issuing a rating that brands have a credibility to preserve and that they wouldn't obtain anything by ripping off customers with a fudged efficiency ranking.

The assumption is made (based on typical use) that the Air Conditioning will certainly be used in about 7-8 hrs a day in March and also 15-16 hours in May and also might be simply 1-2 hrs in August. And that is how annual electricity consumption is calculated."

2. Comprehensive explanation for the method of assigning performance ratings.

A method of computation is defined for every classification of appliance. "Performance is defined as result by input. The input is power consumption for each device, however after that result is different."

The outcome for ac system, as an example, is cooling down ability; for fans, it's air delivery, and so forth.

So for every home appliance an approach of estimation is defined. The lower 10 percent has actually to be eliminated from the market as well as then remainder 90 percent has to be separated right into 5 (different) star scores.

3. Buying an extra energy efficient geyser might have a big impact on your power costs

Yes, it's true that the Company does not examine each device before providing a score yet the really feels that brand names have a credibility to preserve and that they wouldn't get anything by cheating clients with a fudged effectiveness ranking. Nevertheless, the does obtain specific appliances' energy efficiency inspected and also sets up notifications on its web site if they deliver reduced efficiency than what their star tags recommend.

If customers discover that the model's performance is not as per the specs in the label, then the consumers can submit a composed grievance to SEER. In such cases, SEER once more checks the models on its own and also if the examination results do not match the label specs, then the version sheds ranking and producer needs to remove the item from the market."

4. Why should you search for star labels when buying devices?

A greater star ranking indicates more a much more efficient appliance, which implies a reduced electrical energy bill every month. If the use of any type of appliance is high, one need to go for a high power effective appliance.

A higher star rating implies even more a more effective device, which means a lower electrical energy costs every month

Ceiling fans have a very high usage of 10-20 hrs per day and cumulatively all fans can contribute to approximately 4-5 systems each day in a family. Fridge as well as water heaters once again can add to up to 3-4 units each day in any type of residence. Looking at effective appliances for such appliances make a whole load of sense.

5. Should you simply purchase the highest star-rated appliance in each category?

No, this would be a mistake. Star rankings are a good indication of a home appliance's power efficiency but the system is not without its problems. Here are a few of the things to keep an eye out for before buying an appliance based upon the label:

5.2.1. When was the device ranked?

Each star tag also mentions the year in which the home appliance was ranked. Since the SEER keeps updating the star label system, it's best to buy a home appliance that was ranked lately.

5.2.2. Not every classification of home appliances gets a star tag.

SEER star rankings are not necessary for ceiling fans as well as suppliers may minimize the output for a higher score. If the fan doesn't properly spread air around the space on a hot summer day, after that there's no point of having higher power performance. Groups such as multi-door refrigerators don't yet have a star scores system, so the choice becomes tough.

5.2.3. Sub-categories of devices have various criteria

The most convenient example for this is a/c. Window a/c unit and also split air conditioners have different star labels, so a five-star ranked window Air Conditioning may actually be consuming even more power than a three-star split Air Conditioner. The label also points out the amount of units each appliance eats each year, which could offer you a much more accurate suggestion of the home appliance's efficiency.

5.2.4. Compute ROI

There's likewise the concern of return on investment (ROI). One of the most power efficient technologies might exist in an item that's way over your spending plan, and in such cases you may be far better off looking for an item with the highest ranking in your budget plan. This varies a great deal from situation to situation, so you're far better off computing the ROI prior to buying. An ineffective window Air Conditioner is most likely to cost you a lot much more in regards to power costs than a very effective inverter Air Conditioner, but not all choices are that simple.

5.2.5. Take a look at the innovation made use of

In many cases, lower effectiveness is the cost to be spent for much better technology. There are specific devices like Televisions, that do have star rankings, but higher electrical energy consumption is in fact credited to innovative modern technologies. In such situations if advanced innovation is a priority, after that star ratings may not be an efficient method to filter out home appliances. As we pointed out in our Air Conditioning acquiring guide, lots of new home appliances utilize the inverter technology to lower power intake. In these situations, the appliance might eat much less power than its non-inverter counterpart.

5.2.6 Why do not all groups of home appliances have rankings?

We did send out in a thorough questionnaire to the SEER however they didn't respond, so the ideal we can do is think. It has been observed that groups where the manufacturers are primarily international, the SEER scores are being updated every 2 years.

5.2.7. Multi-door refrigerators (such as the one on the left) do not yet require SEER rankings.

There are other categories where the modern technology hasn't improved much and also labels are missing or not required. The modern technology enhancement has actually been exceptionally sluggish as well as also the standard is still not obligator.

There are groups such as multi-door refrigerators that still don't have SEER scores, but if the market share of these products increases there's a higher opportunity of them being added to the SEER rating program.

Page intentionally left

What is Smart Home Control System?

Discover why you ought to link every one of your house's electronic devices with a home control system.

A house control system connects all the electronics in your residence. It indicates that not just is your phone and also your tablet connected to the net, yet your thermostat and your lights too. Connection is one of the most important feature of house automation and also it stands for a suggestion referred to as the 'Internet of Points'.

Nowadays there are a lot of off-the-shelf house automation 'solutions,' but without a home control system (additionally referred to as a hub), you mind find yourself with one application for your lights, one more for your home heating as well as yet one more for your enjoyment. This degree of issue beats the initial objective of home automation by making points more difficult, instead of much less.

1. Exactly how an automated residence could change everything.

Net connected devices, occasionally known as clever gadgets, have three significant advantages over their traditional equivalents. These are benefit, efficiency as well as control. Smart devices are convenient because, as the term home automation recommends, specific features can be automated based upon your way of living, including your house's lighting, heating and also safety and security. Some devices, such as Google's much hyped Nest Understanding Thermostat, also assert that they can pick up from your practices as well as react as necessary.

Smart tools give you regulate by permitting a much more advanced level of communication with your devices. You can from another location control your devices through an app on your phone, allowing you to turn on the home heating before you also get house from work.

2. Why you require a house control system.

When they are connected to each other through one central gadget, your connected devices function best. This is where the residence control system can be found in. A house control system acts as a hub that allows you to interact with all of your devices via one instinctive interface, normally an application. It's valuable, some might say essential, for every one of your tools make use of the exact same networking procedures, whether that be Wi-Fi, Bluetooth or Z-Wave. If your tools don't make use of the exact same protocol, after that handling them may need a variety of different interfaces, which detracts from the ease of residence automation-- defeating the objective rather. It may be time to reconsider if it's not going to offer you terrific control, benefit, or effectiveness.

It's possible to begin little by automating simply one room, or installing a small residence theatre system. But there's always the opportunity to upgrade to whole residence automation whenever you really feel that the moment is right. Just like a mobile phone, a clever residence is an experience that you'll soon ask yourself just how you ever lived without. Our intricate installment jobs are limited in extent only by your imagination.

A residence control system attaches all the electronics in your home. Smart tools are hassle-free since, as the term residence automation recommends, specific features can be automated based on your way of living, including your residence's home heating, lighting and safety and security. You can remotely regulate your devices via an application on your phone, permitting you to transform on the home heating before you even get residence from job. A residence control system acts as a hub that allows you to communicate with all of your devices through one user-friendly user interface, commonly an application. If your gadgets don't use the exact same protocol, then managing them might require a number of various interfaces, which takes away from the ease of house automation-- defeating the function rather.

3. Smart Residence Tools That Automate Your Home

Smart-home Technology

Your residence is your castle, and a castle must be safeguarded in any way times. Certainly, you're not always in your home. You have actually got to most likely to function as well as you require trips to escape work!

So exactly how can you maintain a watchful eye over your house as well as ensure your belongings are secure when you're away? 2 words: home automation. There are lots of companies reinventing the house automation sector.

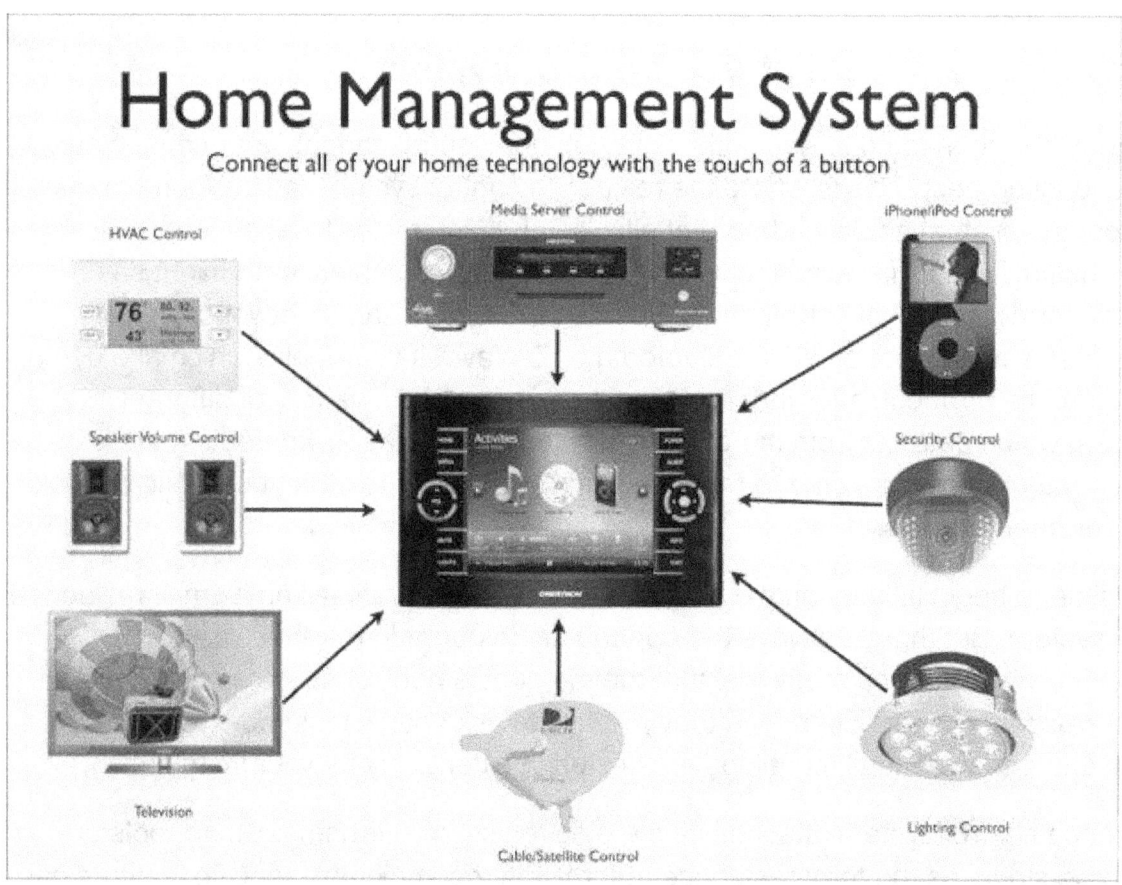

Residence automation has been around for a long period of time, yet the appearance of mobile phones as well as Wi-Fi has actually made the technology much more convenient as well as inexpensive. It may sound like science fiction, however you can now manage all of the devices in your home right from your mobile phone.

We still have not gotten to the day where the recipes do themselves, yet these 9 clever residence tools can make your life easier.

3.1. Smart house safety system

Have you ever wanted you could watch on your home when you run out town? With automated protection innovation such as Piper, you can now check your house security system 24/7 from your smartphone with a wise residence safety application. If any kind of unusual movement is found, Piper gives you a scenic sight of your house as well as should alert you.

If your house is intended to be empty throughout your household trip, Piper ought to inform you with a phone call, message, or e-mail if, as an example, a home window is opened or it discovers a loud noise on your building. You can even see a video of what caught its focus on your phone.

3.2. Video buzzer cam

When you're on your next exotic holiday, not every person will certainly know that you're away from home. Ring's VideoDoorbell is a Wi-Fi enabled doorbell camera that enables you to video clip conversation with visitors through your smartphone.

Have you ever before missed a vital package since you were simply a couple of mins late getting home? With doorbell video cameras, you can let the distribution man understand that it's secure to leave the shipment at your door.

3.3. Smart plug as well as clever electrical outlets

With the Belkin WeMo wise power outlet, you can utilize your house Wi-Fi network to transform tools on as well as off right from a smart home app. This house automation technology can even tell you just how much power or energy you make use of on your electronic devices, offering you useful understanding right into which home appliances as well as electronics are costing you the most each month.

3.4. Smart thermostat

When you leave your house can be a big waste of cash, failing to remember to change the thermostat. Smart thermostats are making this circumstance a distant memory. With technology like Nest, convenient control of your home's thermostat is currently right at your fingertips wherever you are.

Best of all, its innovative automation modern technology consistently finds out to pre-heat/cool your house based on your habits so that you are comfortable at all times. With a Wi-Fi enabled thermostat, you can take complete control of your residence's temperature level settings.

Your thermostat isn't the only device you can regulate from a clever house application. Here is a listing of the most effective clever devices to aid you proceed the modernization of your house.

3.5. Smart irrigation controller

One easy thing to forget when you take place getaway (besides your toothbrush) is locating someone to sprinkle your plants. There is also residence automation technology that has your yard covered. Whether you are searching for complete Wi-Fi watering systems, dampness sensing units, wise gardening apps as well as cloud platforms or gardening Do It Yourself and open source systems, plant watering sensor systems can assist.

3.6. Smart smoke and carbon monoxide detectors

Fire and also carbon monoxide are two of the most significant safety and security risks in your home so it's reassuring to have the capacity to monitor them from anywhere with a wise house application. Nest also created an alarm system that syncs with your smart device to alert you if the smoke or carbon monoxide alarm system goes off.

3.7. Smart fish tank controller

Beyond your home as well as garden, home automation modern technology can even shield an additional valuable belonging. Fish tank conditions are provides you complete control over your tank from your smart device. From ozone control to automated feeding, this tool eliminates the requirement for a Nemo-sitter on your next holiday and takes the marine biology uncertainty out of your aquarium.

3.8. Smart locks.

Do you ever leave house and fail to remember whether you secured the back door?

The wise deadbolt offers you a virtual spare key so you do not require to hide anything under your doormat. With this home automation technology, you can control your locks from anywhere with your mobile gadget.

3.9. Smart residence

What if you were able to manage several tools in your house with one solitary application? It doesn't quit there; the smart system is a system that allows the multi-product connection, allowing your wise gadgets to speak the very same wireless language.

These 9 groundbreaking residence automation innovations can be used to protect your residence straight from your smartphone. While wise home applications might help make your house much more effective or protected, they don't safeguard your residential or commercial property or items if something unforeseen takes place.

Page intentionally left

How to select CCTV System?

Setting up any closed-circuit television (CCTV) safety system calls for four fundamental elements: video cameras, a recording system, a display, and video clip monitoring software application. A variety of supporting equipment is also needed to become fully practical, and that consists of power line, though the exact requirements rely on the primary CCTV's parts.

Listed below we'll examine the major equipment categories, and also what to take into consideration when searching for each.

1. Consider Cameras

If you're developing a CCTV video camera system, you have two electronic camera choices: IP (net protocol), or analog. The previous is the recommended selection for any person thinking about contemporary capacities, as it is the best for the large majority of applications. As a matter of fact, you ought to just go with an analog camera if you have an existing digital recording system that needs older innovation.

2. Both Basic Sorts Of CCTV Equipments: NVR and DVR

When it involves video clip recorders, you likewise have a number of alternatives: NVR (network video recorder) as well as DVR (electronic video clip recorder). Although DVR's prevail in American families as well as likely much more acquainted to the ordinary consumer, NVR is really the alternative you should opt for due to its combination with IP electronic cameras. Furthermore, NVR's are commonly a lot more expensive than its counterpart, however the ability to collaborate with high-resolution IP video cameras surpasses the prices in most situations.

When incorporated, IP and also NVR innovation make your CCTV system simple, adaptable, as well as future-proofed. With each other, they give wireless capacities, supply superior total picture high quality, and call for less cable televisions. By contrast, analog cams call for not one, yet two cables each, minimum, are lower resolution, and harder to mount.

When given the option, this is one area where you get what you pay for, so pick the much better system. If you wish to update an existing DVR system, you can transition to a crossbreed system, though it can be complicated to obtain everything effectively interacting. Once you have actually decided on your high-level tech, it's time to select a monitor and supporting tools.

3. Grab a Monitor

A monitor lets you carry out perhaps one of the most vital function of a security electronic camera-- in fact seeing the video. Certainly, your selected monitor will mainly be dependent on your electronic camera and also recording software program.

For instance, if you have 1080p or 4k resolution cams installed, only a high-definition screen will certainly allow you take advantage of the picture-perfect detail the video cameras provide. On the other hand, if you have older or lower-quality elements, you can conserve a couple of dollars.

It's likewise worth thinking about the number of cams you'll be installing for your company, and how many spaces you require to be monitored. If you're under the perception that you'll call for dozens of monitors, straight out of an activity motion picture, that may be overkill. The only validation for such a set up would certainly be if you operate a large facility or big substance and require to view every space and also cranny. 3 to 5 displays is a great beginning factor, and also you can modulate up or below there based upon your requirements.

4. Handle Your Video With VMS

Once you choose the cam, taping technology, as well as display arrangement that's best for your organization, you're 90 percent of the way toward producing a strong CCTV system-- however you're not done yet. One other essential component to take into consideration is your VMS (video monitoring software program). This is the means by which a human can regulate the NVR or DVR videotaping system.

You have actually most likely used a fundamental DVR in the house when recording your favorite TELEVISION shows. Obviously, business-grade protection options remain in a different course, but they also aid you discover, separate, and essence one of the most vital moments in a recording.

You'll have to do a bit of study to find out which VMS is best for your situation. Bear in mind that your choices will differ based upon your original choice to select NVR or DVR modern technology. Just see to it the software program is compatible with your system, and also review exactly how easy to use the user interfaces are some will take even more training than others prior to you become an experienced customer.

Support Your System with Cables, Power Products, as well as Routers

Similar to the abovementioned attributes, depending upon your system, you'll call for a range of sustaining modern technology to connect the dots as well as guarantee seamless assimilation. Analog electronic cameras, for example, call for a coax cable to connect to a DVR, as well as an additional power cable. On the other hand, IP choices can attach to the NVR recording center and get PoE (Power over Ethernet) from one cable. Wireless systems also call for a router, while wired versions do not.

Cameras are one of the most Crucial Component

At the end of the day, the initial decision regarding what electronic camera and also recording parts to utilize is the most crucial as well as will certainly have a massive effect on the type of sustaining technology to successfully run your system. If you can manage it, go with NVR recording and IP electronic cameras, but if you have an older, existing system, you can add analog cams or upgrade to a hybrid.

Once the cam and also recording tools are chosen, research several monitors, get a video clip management system as well as purchase the cable televisions and also devices you require to connect whatever.

Just how to determine Electrical Mistakes

It's irritating if an electrical outlet or home appliance doesn't work. However with a little bit of sensible investigator work, you should be able to discover the root cause of the issue - and, in many cases, fix the fault yourself. Prior to you begin any type of type of electrical work, though, you'll require to recognize how to separate a circuit and also double-check that it's dead.

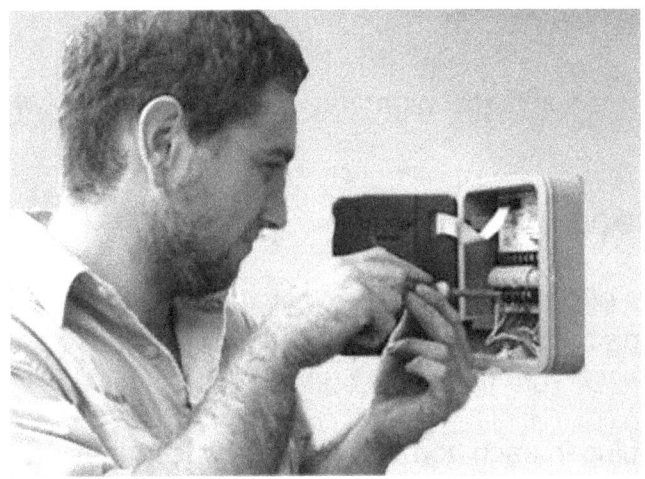

Increased, Safety and security

For your safety and security, electric items need to be set up according to local Building Regulations. If in any type of doubt, or where required by the regulation, get in touch with a qualified individual who is signed up with an electrical self-certification scheme. Additional info is available online or from your Neighborhood Authority.

Never ever take threats with electrical safety. Prior to you start any kind of sort of electrical job, you need to follow these following security preventative measures:

1. Turn off the major power at the customer unit/fuse box. Isolate the circuit you intend to service by getting rid of the circuit fuse. Put this in your pocket to stay clear of unintended replacement
2. If you can, - Or change off the breaker and also lock it
3. Affix a note to the unit to recommend you are working with the circuit
4. Examine the circuit is dead with a socket tester or voltage tester/meter for lights circuits

Comply with these actions to locate the cause of your electrical issue.

1. A plug-in device doesn't work

If a plug-in light does not work, try transforming the bulb. If it doesn't, try once again on a various power circuit (most likely on one more floor). If it works there, you may have a dead circuit.

If the home appliance doesn't work in an outlet you understand is working, inspect the flex connections in the plug and replace the fuse - making sure it has the appropriate score.

Still not functioning? The appliance might have an inner mistake which requires some specialist focus.

2. A circuit is dead

Switch off all the lights or disconnect all the appliances on the influenced circuit. Switch off the major separating button at the customer unit, as well as repair the circuit fuse or reset the breaker or RCD. Turn the major button back on.

Turn on each light or plug in each home appliance consequently to discover which product on the circuit is creating the fuse to strike or the breaker to tripping. When you locate it, separate the circuit once more and also inspect the fuse, the connections and also the flex (see over).

The mistake may exist in the repaired circuitry if the fuse impacts or the circuit breaker tripping once again. Call a qualified electrical expert.

3. A wall surface or ceiling light doesn't work

Inspect if the various other lights on the circuit are working. Otherwise, follow the actions for a dead circuit below.

Transform off the light at the button and replace the light bulb if the other lights on the circuit are working.

If that doesn't help, turn off the power and also inspect the cable/flex connections at the light. Still with the power off, examine the condition of the flex with a connection tester, as well as replace it if needed.

If that doesn't work, transform off the power once more, get rid of the switch cover and inspect the cable connections. If they're fine, attempt replacing the button.

Still not working? Call a certified electrician.

4. All circuits are dead

If the circuits in your home are protected by an RCD (residual load-current device), check to see if it has stumbled. If so, reset it. If it tripping once more, carry out the look for defective lights, devices and a dead circuit. If the problem continues, call a certified electrical contractor.

If the power to the area has been reduced, check with neighbors or your electrical energy vendor to locate out. Otherwise - and also you can not discover a trouble with your domestic circuits - contact your electrical energy distributor. They'll have the ability to examine the primary supply cord and also service fuse.

5. Expanded, Fuses and circuit breakers

Each of the electrical circuits in your house is provided with a fuse or a circuit breaker. They can be either an MCB (mini circuit breaker) or RCBO (recurring existing breaker with overload protection).

These gadgets safeguard the circuit against overwhelming, which can generate heat within the wiring that melts the insulation and triggers a fire.

They additionally respond to short circuits that are created when the current-carrying cores of wires enter into contact with each other. This can occur if the cores end up being loose inside an electric device, or if the cord is pierced accidentally by a drill or nail.

Fuses have an unique cord that thaws, cuts and also divides off the flow of electrical power if the circuit draws too much existing or a short circuit happens.

The cord could be exposed within the carrier, or it can be included within an unique cartridge. Breaker are tripping switches over that turn themselves off under the same situations, and also can be reset by pressing a button or operating the button.

The demand positioned on circuits varies (light fittings take in less electricity than the majority of plug-in devices, as an example). As well as having actually different sized wire, the circuits are secured by integrates or circuit breakers with different rankings.

Illumination circuits are protected by 5 or 6 amp integrates, outlet circuits by 30 or 32 amp integrates, an immersion heating system by a 15 or 16 amp fuse, and so forth. It's very vital that you make use of integrates of the right score. One with as well reduced a ranking will certainly keep blowing, while one with a rating that's expensive could not secure the circuit against straining - with possibly fatal repercussions.

5.1. Safety first - Hassle RCD stumbling

To prevent your RCD-protected systems from stumbling, you should collaborate with the whole power supply turned off. Turning off a breaker or getting rid of a circuit fuse only isolates the L (online) side of the circuit, while the N (neutral) remains attached to the mains. This is fairly risk-free for working with the circuit, however it implies that any type of contact with the N cable will certainly cause the RCD to trip and also switch off the entire home supply. This is not just bothersome, however can additionally be dangerous when you remain in the center of a repair service.

a) Expanded, Exactly how to alter a cartridge fuse

Cartridge integrates are easy to replace - however see to it you use the correct fuse score for the circuit, as merges vary in dimension and also colour coding according to their score.

Zoom: [photo description] Step 1.

Turn off the power and also eliminate the cartridge fuse. Some are just kept in springtime clips as well as can be prised out, while on others you'll need to open the fuse provider by releasing a screw.

Zoom: [picture description] Step 2

Press a brand-new fuse of the right rating into the spring clips, or insert it into the open ends of the provider's pins. After that, if essential, reconstruct the fuse service provider. Check the main power switch is off and also replace the fuse - typically, the pins are balanced out to one side so it'll just fit one means round. After that, you can recover the power.

b) Broadened, How to alter a fuse cable

You must be able to see that the fuse cord has thawed if a rewirable fuse has blown. You must replace it with brand-new fuse cable of the appropriate amperage.

Zoom: [photo description] Leading tip - Connection unit fuses

Taken care of home appliances are completely wired to a connection device that has an on/off wall surface button and a cartridge fuse. To replace a connection device fuse, first turn off the connection system button. Unscrew or prise out the fuse holder. Get rid of the fuse and also put a new one of the appropriate ranking. Make certain you press or screw the holder back fully before you restore the power.

Zoom: [picture summary] Step 1

Shut off the power, raise the fuse cover on the consumer system as well as remove the blown fuse. Release the incurable screws as well as get rid of the fuse wire.

Step 2

Take a size of fuse cord of the appropriate rating as well as insert it in the carrier. Examine the main power button is off, after that refit your fuse.

Page intentionally left

Top 10 electrical Tools for Home essentials

If you're an expert electrical expert you're most likely looking for the latest and also biggest tools that can supply excellent outcomes fast for all your electrical tasks. The majority of devices are conveniently available, however if you're in the electrical trade, you recognize that there are some high quality branded devices that can last for years and also other more affordable tools commonly known as 'throw out devices' that need to be replaced over and over once more.

More often than not quality constantly dominates price and selecting the right devices wisely, even if you're simply beginning or want to construct an electric package for home usage will save you much time as well as headache in the long run. In this post we'll assist you choose a few of the leading tools that every residence customer, beginner or pro electrician needs in their tool-bag:

1. Measuring tape

An essential for novices as well as pro electrical contractors alike, a tape measure is necessary for gauging heights for button and also outlet positioning, focusing illumination components and so much more. For under $15 we recommend branded self-centering tape, clear-coated blade defense for even more long lasting markings as well as designed to fit easily in hand.

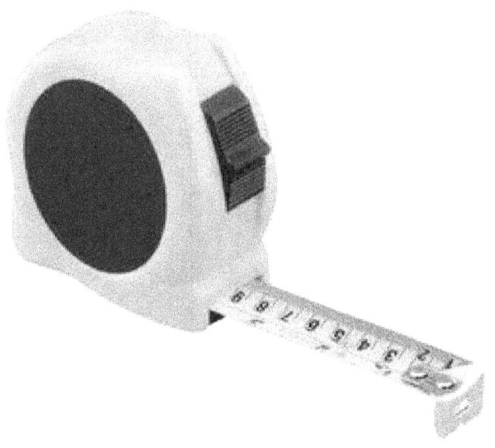

2. Multimeter

A multimeter is an important all-in-one tester made use of to determine voltages, existing and also resistance in an electrical circuit and can aid you situate power fluctuation triggers like inferior circuitry. Branded equipment offers some spending plan and also pro multimeters to match your requirements.

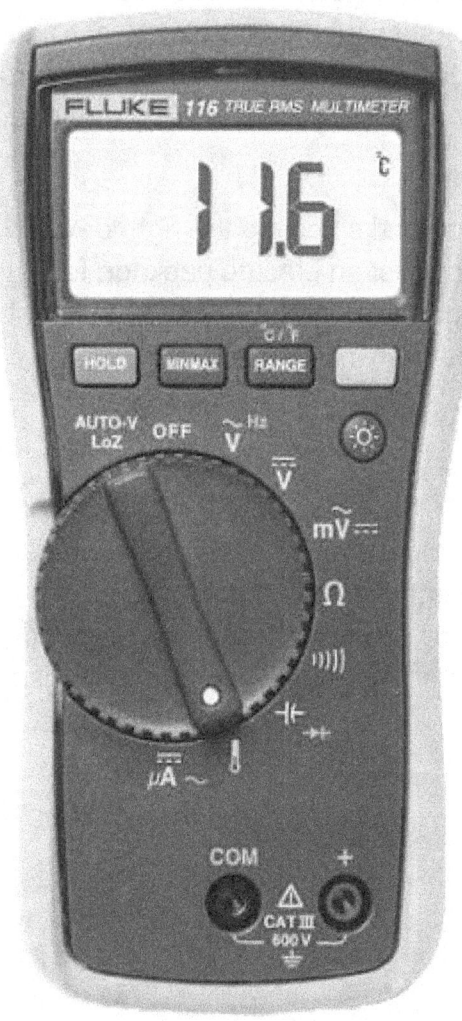

3. Cable stripper

A handy tool to strip or cut off the insulation on cables. A good top quality cable pole dancer will do a tidy work every time and is developed with a cutoff part as well as different sized cutting teeth for various sized cords or wire.

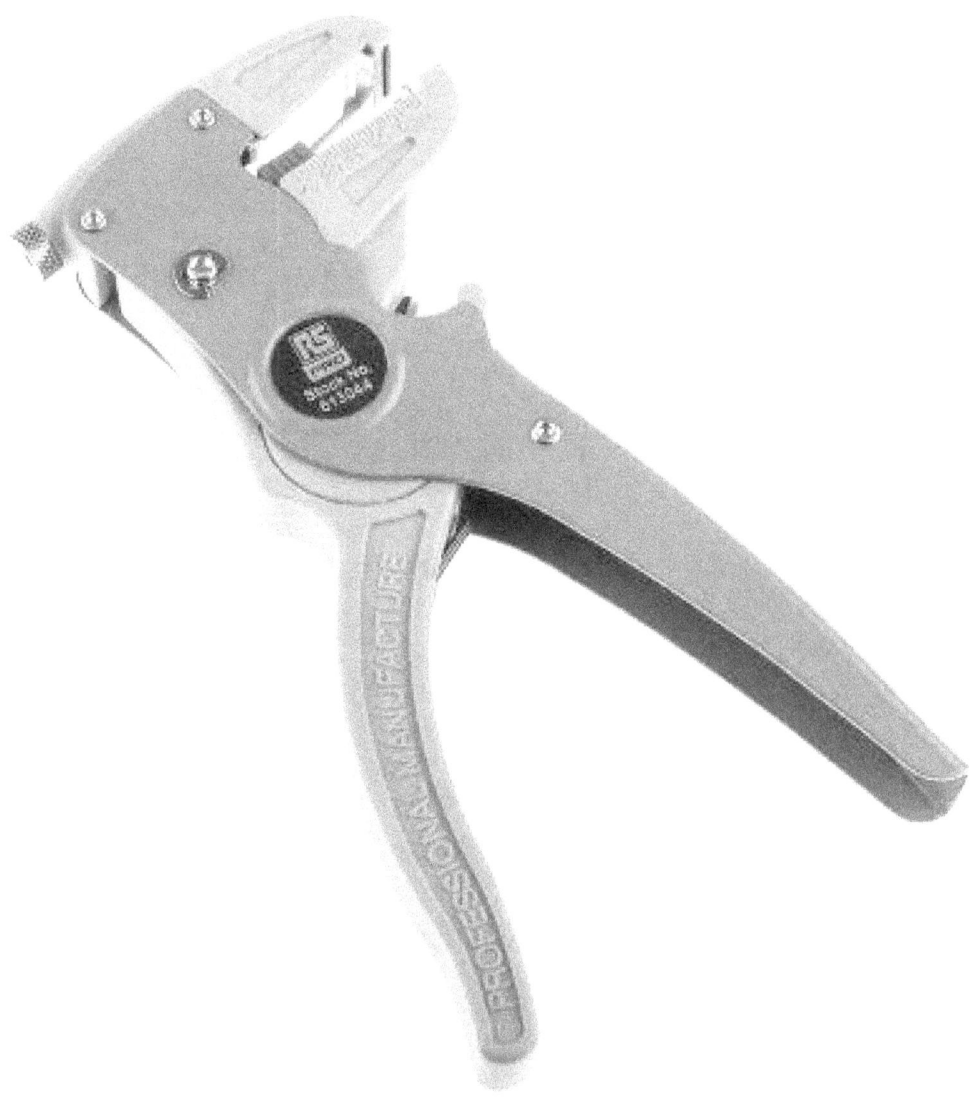

4. Fish tape

Likewise called draw wire/tape, fish tape is made use of by electrical experts to path brand-new circuitry through walls, metal, electrical as well as pvc avenue. We suggest wide steel fish tape by branded devices-- readily available in different lengths together with branded wire lube to assist in directing the tape with slim spaces.

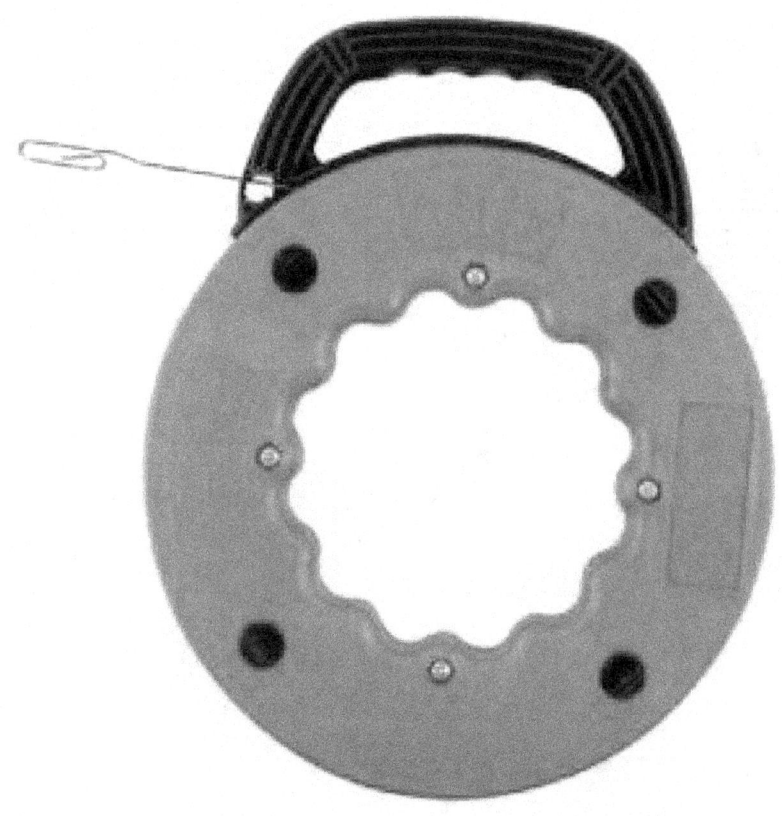

5. Non-contact voltage detector

There'll be lot of times you'll need to do a quick safety check to see if there is an existing present or a circuit is certainly live. A voltage detector is either automated or has an on/off switch. Select the one that suits your spending plan and demands. We suggest the branded voltage tester; a sturdy tester that will certainly supply accurate test analyses as well as a lengthy service life on the job-site, in the shop and at home. At under $20 we advise the branded equipment ncvt-2 dual array non-contact voltage tester.

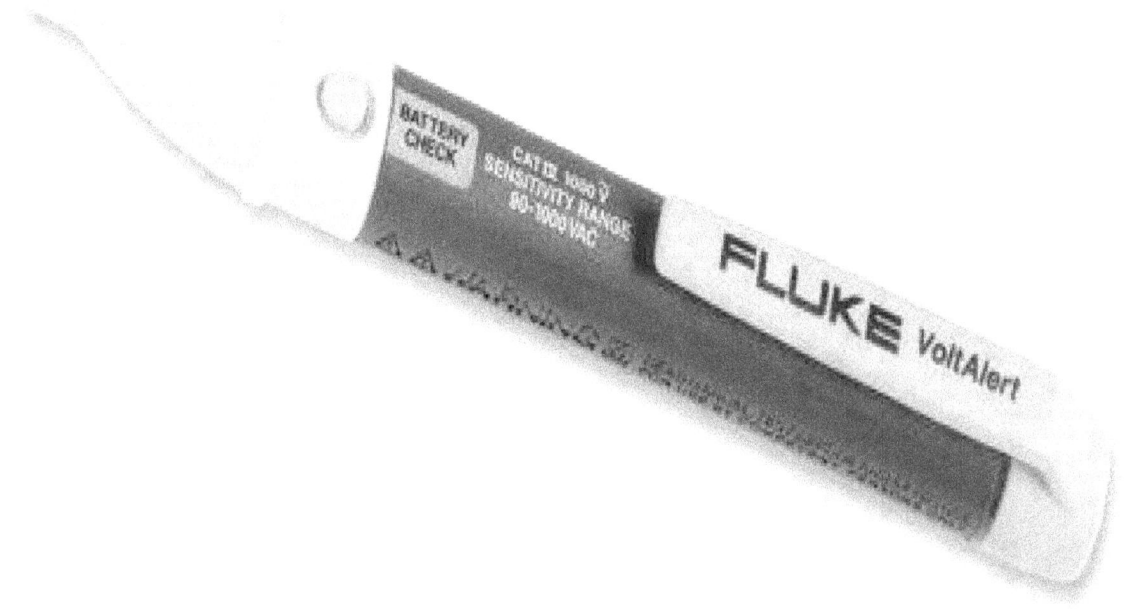

6. Plier

Pliers come in lots of kinds depending on what requires to done. You'll likewise need a high quality collection of do-it-all pliers. These can cut cord, spin cords with each other utilizing their squared off suggestion and hold and draw wire.

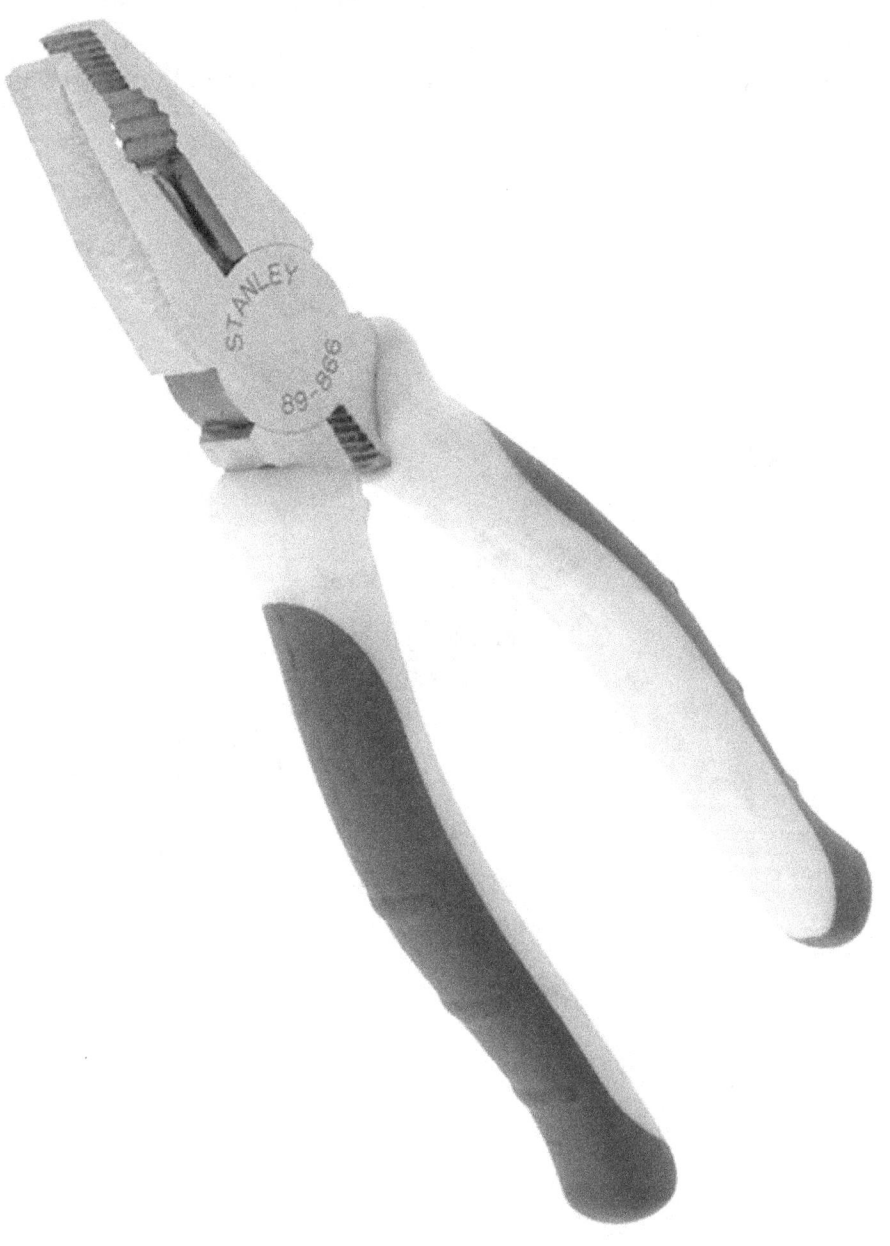

7. Level

A great installation starts with getting the fundamentals right. A degree is utilized to see to it all your job is level including straight electrical outlet covers, wall plates and also buttons. For simply over $5 the hands-free magnetic torpedo degree is a steal. Portable and also lightweight, it gives precise analyses, including horizontal, upright and at 45-degree angles.

8. Flashlight

Any pro electrical contractor will certainly inform you that appropriate lighting is vital to any kind of electric work, which you should never get to as well as attempt right into a panel without appropriate lights. When illumination problems on duty site are not the very best, a good flashlight or job light can save the day. We offer over 50 kinds of flashlights to meet any type of lighting need in any space.

9. Cable crimper

Electrical problems are typically difficult to trace down and also quite often the concern may be recurring and also the outcome of a poor link. A cable crimp tool is among the most effective ways to repair wires creating damaged circuit connections. Purchasing a top quality crimp tool makes certain a long-lasting seal and also will certainly give you years of trusted usage.

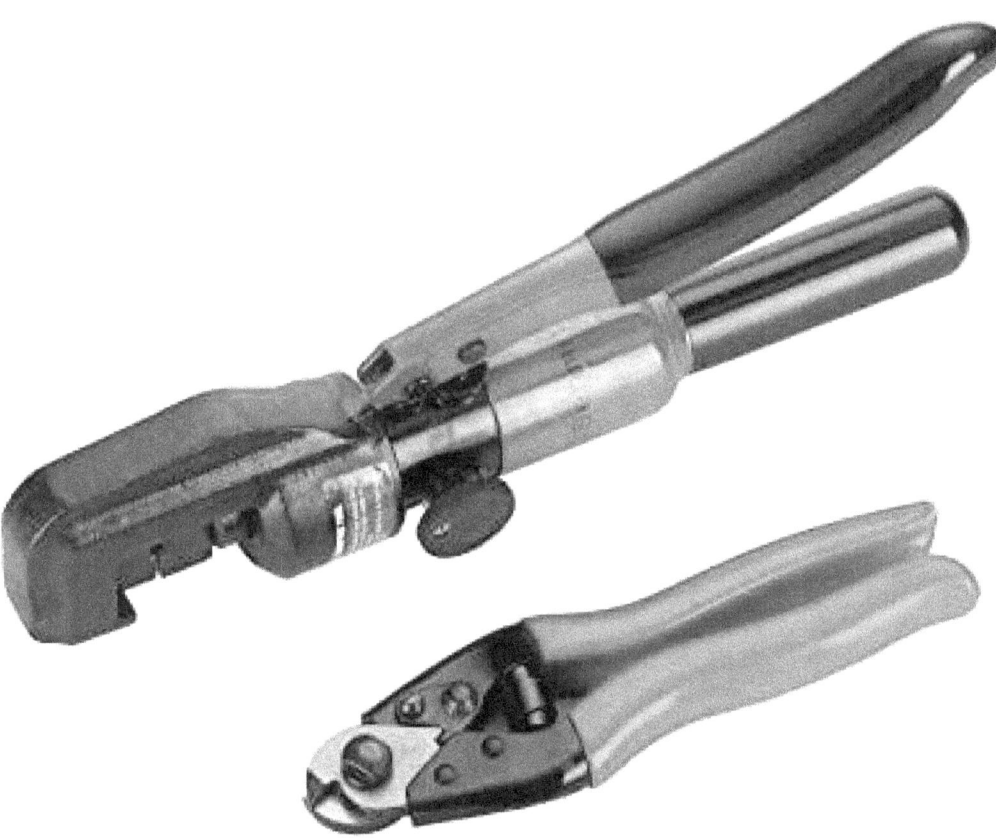

10. Screwdrivers

A quality set of screwdrivers can last for life. You'll need various types like screwdriver, head screws, as well as a collection of straight blade screwdrivers. There are a lots of options readily available consisting of digital screwdrivers, magnetic screwdrivers for better hold, mutli-tip screwdrivers with compatible tips, accuracy screwdriver collections, pocket clip-style screwdrivers and even more.

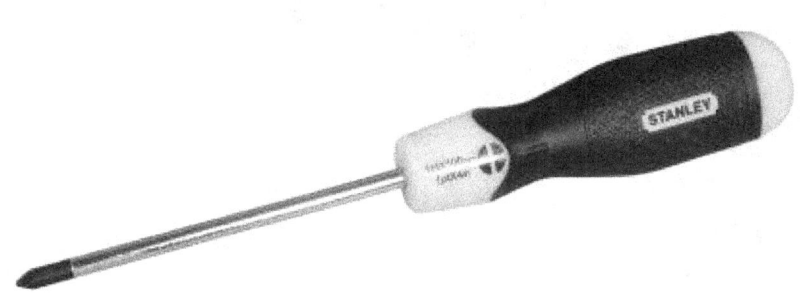

How to handle the Multimeter?

Basically, a multimeter is an electronic device that determines existing, voltage, and resistance. (You might also have heard them referred to as multi-meters).

Current measured in amps, resistance is measured in ohms, and also voltage is determined in volts.

There are two major types of multimeters: analog meters and also digital meters. Though analog meters use a needle to make dimensions, today most people use digital multimeters.

Digital multimeters are typically much more accurate and also use more consistent readings.

Nowadays, multimeters are utilized for great deals of different objectives. Some hackers have even found methods to transform them into clocks!

1. Four Main parts of the multimeter

Before we get involved in more detail regarding exactly how to read multimeter icons, initially we wish to make sure you know the 4 main parts of the multimeter itself. They are:.

a) The Display:

This is the screen where your measurements will certainly be shown.

b) Buttons:

Depending on the kind of multimeter you choose, you'll have various options and positioning.

c) Dial/Rotary Change: This is where you select what system of measurement you require.

d) Input Jacks/Ports: These are the locations that you put your examination leads within.Test leads are insulated wires that perform the multimeter to the item you're measuring as well as checking.

Bear in mind that matters and also figures tell you the resolution of your multimeter.

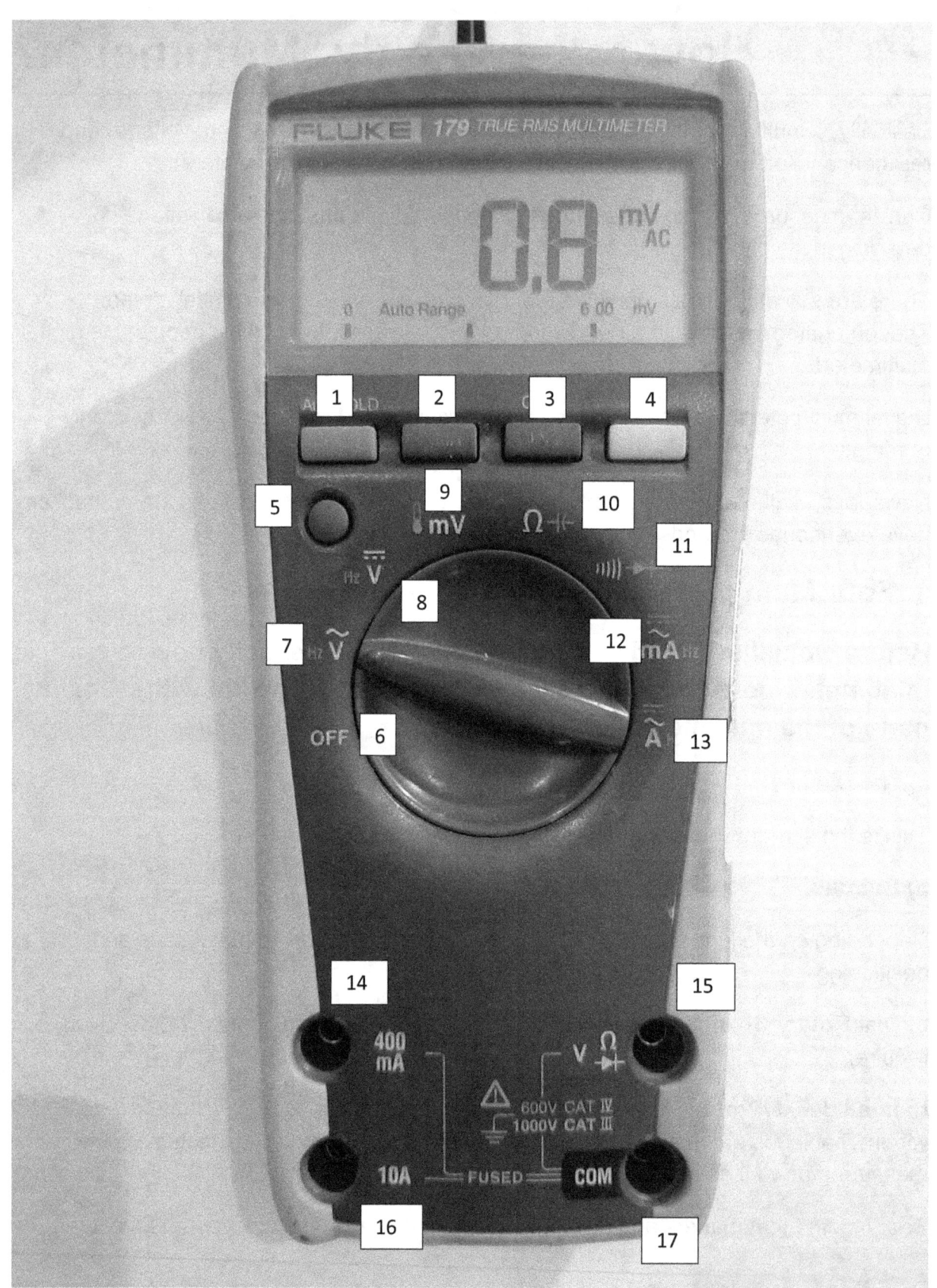

2. How Can I Read My Multimeter Signs?

Fluke Multimeters are just one of the most prominent brand names of multimeters made use of today. The good news is, their signs are additionally the requirement for nearly all various other multimeters. So, we'll be assisting you to check out the most common symbols of multimeters, a number of which are seen on the Fluke brand. Once the concept is cleared then we can choose any brands which is good in market.

Keep in mind: On some multimeter versions, you'll discover that there are extra, yellow icons around your rotary switch/dial. Make certain you press the Change Switch if you require to accessibility and also check out these. This is similar to you would on a computer keyboard.

Button 1:

The Hold Switch. Usually situated in the leading left-hand edge of multimeters, this switch locks your meter reading/measurement into place after you have actually taken it.

This is particularly helpful if you're doing a project that needs you to keep a specific dimension within your reaches. It's likewise a terrific attribute if, throughout a testing of the probes, you can't review your multimeter completely.

Button 2:

The button two is used for check the minimum, maximum and average of measured current / voltage readings

Button 3:

Range Switch. This switch will typically be discovered throughout the top of your multimeter, as well as has a "Lo/Hi" sign over it. This will aid you "click with" different meter ranges.

While today, a fantastic bulk of multimeters have auto-ranging, you can also select a details array on some versions-- like switching from Ohms to mega-Ohms.

Button 4:

Shift: Herz. This is usually the shifted analysis above the AC Voltage choice, marked "Hz." This multimeter symbol will certainly inform you your circuit's or devices's frequency.

Given that a lot of will operate at either a variable or fixed regularity, you need to make sure you know which one you'll be collaborating with before beginning your measurement.

Button 5:

Brightness Sign. Similar to on an iPhone, this is the button that can allow you darken or lighten your screen, making it much easier to read if you're taking measurements outdoors.

It's recognized by-- you thought it!-- a little illustration of the sun.

Button 6:

Button 6 is the rest position of the multimeter as well as switch off device.

Button 7:

A/C Voltage. This multimeter sign is recognized as a resources "V" with a curly line above it that looks a bit like an accent mark in the Spanish language.

Likely, this will be the setting you use to take electronic dimensions most often. It gauges the voltage of your things, regardless of the setting or item that you're operating in as well as with.

Generally, you ought to anticipate to see readings in between regarding 0-1000 Volts.

Button 8:

DC Voltage. This button is likewise a single capital "V," as well as it has 3 hyphens (---) over it, after that a solitary straight line in addition to that. It type of appear like a V with a road illustration on top of it!

This is the setting you'll use when you're determining smaller sized circuits, batteries, as well as even indication lights!

Button 9:

DC Voltage. This button is likewise a single capital "mV," as well as it has 3 hyphens (--) over it, after that a solitary straight line in addition to that. It type of appear like a milli Volt with a road illustration on top of it!

Button 10:

Ohms. No, you're not chanting this is a yoga exercise mantra. A minimum of, not when you're reading multimeter icons. This appears like an Omega letter, as well as it helps you to get one of the most exact resistance reading feasible.

Also far better? This button can also assist you to identify whether or not a fuse has actually blown. If your meter displays "OL," after that the fuse blew out, and you can remove it.

As a note, make certain you've taken merges out of the circuit when you're using the ohms setting on your multimeter. Whether you're dealing with your own or with a team, regarding 143 electrical contractors die every year from electrocution. Better risk-free than sorry!

Button 10:

Shift Capacitance. This is usually the shift choice on your Diode Examination button and also appears like two "T" letters encountering each other. This measures your capacitance.

Button 11:

Diode Examination. This has an arrow pointing to the right, with a plus sign right alongside it. As you might have thought, this tells you if you're taking care of good or poor diodes.

Some people use the ohm setting to test them, this is extra exact.

Button 11:

Connection. This button resembles a load of closed-end parenthesis in a row, like the sign that suggests sound.

Possibly that's due to the fact that it in fact puts out a noises itself! When two points have continuity, you'll get a beeping noise. It's a very easy and outstanding way to see if you have any type of open or short circuits.

Button 12 & 13:

Direct Current. This switch has the same attributes as your Alternating Current button (we'll obtain to that in a minute) however gauges Straight Current rather.

Button 14:

Milli amps Current Jack. OK, so this isn't technically a multimeter symbol. Current can be measured by red lead.

Button 15:

Red Jack. This is the other red jack, generally on the right-hand side of your multimeter.

That's because your red jack measures practically every little thing except for current. That indicates it can help read temperature level, responsibility cycle, frequency, voltage, as well as resistance, among others.

Button 16:

Alternating Current. This button is a funding "A" with a squiggly line over it (once more, think of accent marks in Spanish.).

You'll normally need a clamp attachment to execute functions associated with this option, it's an excellent method to make sure you recognize the quantity of load an item is utilizing.

Button 17:

Black Jack. This is jack, generally on the right-hand side of your multimeter.

That's because your black jack measures practically common point for the current, temperature level, responsibility cycle, frequency, voltage, as well as resistance, among others.

Now that you're a specialist on exactly how to find out these icons, do not make a mistake when it concerns your next house improvement task, and quit losing time making returns on poor-quality projects.

Thank you for choosing our book and encouraging us, kindly leave a feedback or suggestion in this mentioned email.

muthsn@gmail.com

Page intentionally blank

Page intentionally Blank

www.ingramcontent.com/pod-product-compliance
Lightning Source LLC
Chambersburg PA
CBHW081510220526
45467CB00010B/2857